New methods for evaluating one-loop corrections of multi-jet production at the Large Hadron Collider

Dissertation

zur Erlangung des akademischen Grades

doctor rerum naturalium

(Dr. rer. nat.)

im Fach Physik

eingereicht an der
Mathematisch-Naturwissenschaftlichen Fakultät I
der Humboldt-Universität zu Berlin

von
Dipl.-Phys. Benedikt Georg Biedermann
geboren am 30.09.1980 in Mainz

Präsident der Humboldt-Universität zu Berlin:
Prof. Dr. Jan-Hendrik Olbertz

Dekan der Mathematisch-Naturwissenschaftlichen Fakultät I:
Prof. Stefan Hecht PhD

Gutachter:
1. Prof. Dr. Peter Uwer
2. Prof. Dr. Jan Plefka
3. Prof. Dr. Stefan Weinzierl

eingereicht am: 13. Juni 2013
Tag der mündlichen Prüfung: 24. September 2013

Bibliografische Information der Deutschen Nationalbibliothek

Die Deutsche Nationalbibliothek verzeichnet diese Publikation in der
Deutschen Nationalbibliografie; detaillierte bibliografische Daten sind
im Internet über http://dnb.d-nb.de abrufbar.

ISBN 978-3-8325-3580-3

Logos Verlag Berlin GmbH
Comeniushof, Gubener Str. 47,
10243 Berlin
Tel.: +49 (0)30 42 85 10 90
Fax: +49 (0)30 42 85 10 92
INTERNET: http://www.logos-verlag.de

Für meine beiden Füchse.

Abstract

The precise prediction of cross sections for processes at the Large Hadron Collider is key for successful data analysis in new physics searches. Quantum chromodynamics (QCD) offers important contributions to the cross sections. At high energies the corresponding hard scattering matrix elements can be computed within perturbation theory. This allows for the relating of non-observable quarks and gluons as constituents of the colliding protons to jets, collimated bunches of particles, and thus observables of perturbative QCD. In this work, the fully numerical evaluation of virtual corrections for multi-jet cross sections at next-to-leading order accuracy in massless QCD is discussed. To this end, recently developed unitarity methods and integrand reduction techniques are employed. One important advantage of this approach is that the one-loop amplitude is efficiently reconstructed from tree-level amplitudes which circumvents the explicit evaluation of individual one-loop Feynman diagrams. The presented method is a modified version of D-dimensional generalised unitarity with the feature that the one-loop amplitude in QCD with massless quarks is computed entirely in four space-time dimensions. While this is well established for the so-called cut-constructible part of the amplitude, it is a non-trivial aspect for the additional rational contributions. The latter arise as a consequence of the ultraviolet behaviour of the theory and, using dimensional regularisation, they generally require a computation beyond four dimensions. In our approach we incorporate the additional degrees of freedom from dimensional regularisation in a uniform mass shift of the particles circulating in the loop. The rational terms are then obtained in a separate computation, replacing the gluons within the loop with massive scalar particles and the massless quarks with massive quarks. The outlined methods have been successfully implemented in the publicly available computer programs NGLUON and NJET. NGLUON allows the computation of so-called primitive amplitudes — gauge invariant building blocks of the full one-loop QCD amplitude — with arbitrary multiplicity and helicity of the external partons. NJET provides all colour and helicity summed virtual corrections at next-to-leading order contributing to two-jet, three-jet, four-jet and five-jet production. For low multiplicities, the implementation is validated against a large number of known results from literature. Beyond that a detailed study with respect to numerical accuracy as well as runtime performance is being presented. NJET has been applied to compute four-jet production at next-to-leading order at a centre of mass energy of $\sqrt{s} = 8$ TeV. At next-to-leading order large negative corrections of the order of 45% to the leading order cross section and a significant reduction of the scale dependence on the unphysical factorisation and renormalisation scale could be observed. Significant insights as to the origin of the large corrections are further discussed.

Zusammenfassung

Die präzise Vorhersage von Wirkungsquerschnitten für Prozesse am Large Hadron Collider ist entscheidend für eine erfolgreiche Datenanalyse auf der Suche nach neuer Physik. Einen numerisch großen Beitrag zu den Wirkungsquerschnitten liefert die Quanten Chromodynamik (QCD). Die entsprechenden harten Streumatrixelemente können bei hohen Energien im Rahmen der Störungstheorie berechnet werden. Dies ermöglicht, die nicht beobachtbaren Quarks und Gluonen, welche Konstituenten der kollidierenden Protonen sind, mit Jets in Verbindung zu setzen. Jets sind kollimierte Bündel mehrer Teilchen und daher Observable der perturbativen QCD. In der vorliegenden Arbeit wird die numerische Berechnung von virtuellen Korrekturen zu Multi-Jet Wirkungsquerschnitten in nächstführender Ordnung QCD mit masselosen Quarks diskutiert. Dazu werden die kürzlich entwickelten Unitaritätsmethoden und Integrandreduktionstechniken verwendet, deren großer Vorteil darin besteht, dass die Ein-Schleifen Amplitude mittels Baumamplituden konstruiert wird. Diese Herangehensweise umgeht die explizite Berechnung von einzelnen Ein-Schleifen Feynman Diagrammen. Die vorgestellte Methode ist eine modifizierte Version von D-dimensionaler generalisierter Unitarität mit der Eigenschaft, dass die Ein-Schleifen Amplitude in masseloser QCD vollständig in vier Raumzeit Dimensionen berechnet wird. Während dies für den sogenannten cut-constructiblen Anteil der Amplitude wohl etabliert ist, ist dies für die zusätzlichen rationalen Anteile nicht-trivial. Diese treten als Folge des ultravioletten Verhaltens der Theorie auf und verlangen in dimensionaler Regularisierung im Allgemeinen eine Berechnung jenseits von vier Dimensionen. In der hier diskutierten Herangehensweise werden die zusätzlichen Freiheitsgrade der dimensionalen Regularisierung in Form einer einheitlichen Massenverschiebung der Teilchen in der Schleife berücksichtigt. Die rationalen Terme erhält man dann in einer separaten Rechnung, indem man die Gluonen in der Schleife durch massive Skalare und die masselosen Quarks durch massive Quarks ersetzt. Die skizzierten Methoden sind alle in den öffentlich verfügbaren Programmen NGLUON und NJET implementiert worden. NGLUON ermöglicht die Berechnung von sogenannten Primitivamplituden — eichinvarianten Unterbausteinen der vollen Ein-Schleifen QCD Amplitude — mit beliebiger Multiplizität und Helizität der äußeren Partonen. NJET liefert alle farb- und helizitätssummierten virtuellen Streumatrixelemente in nächstführender Ordnung, die zu Zwei-Jet, Drei-Jet, Vier-Jet und Fünf-Jet Produktion beitragen. Für niedrige Multiplizitäten wurde die Implementierung für eine große Anzahl bekannter Resultate aus der Literatur verifiziert. Darüber hinaus werden detaillierte Studien in Bezug auf numerische Genauigkeit und Laufzeit vorgestellt. NJET wurde angewandt, um Vier-Jet Produktion in nächstführender Ordnung bei einer Schwerpunktsenergie von $\sqrt{s} = 8$ TeV zu berechnen. Die Rechnung liefert in nächstführender Ordnung eine signifikante Reduktion der Skalenabhängigkeit von der unphysikalischen Faktorisierungs- und Renormierungsskala, und große negative Korrekturen in der Größenordnung von 45% zum Wirkungsquerschnitt in führender Ordnung. Eine detaillierte Analyse des Ursprungs dieser großen Korrekturen wird ebenfalls präsentiert.

Contents

Introduction

The primary aim of the Large Hadron Collider (LHC) is to investigate the nature of electroweak symmetry breaking and the search for new physics at the energy scale of Tera electron Volt. For a successful analysis of the vast amount of data, precise predictions for the processes within the standard model and its possible extensions are crucial. One of the main tools in collider physics to make quantitative predictions is perturbation theory, an expansion of the hard scattering matrix elements in the coupling constants of the quantum fields. Quantum Chromodynamics (QCD) represents in this context a dominant contribution. First, because the gluons, the major part of the parton flux in the colliding hadrons at high energies, couple exclusively via QCD interactions and, second, because the relatively large strong coupling constant α_s can lead to numerically large values of the hard scattering matrix elements. Although computations within perturbative QCD involve non-observable quarks and gluons as asymptotic states, these partons can be related to jets, collimated bunches of hadronic particles which, in turn, can be seen in the detector. Among the various production modes for jets at hadron colliders, phenomenologically important is pure QCD jet production where every observed jet originates from the same hard scattering matrix element involving exclusively QCD interactions. Such processes are of great interest not only as a background process for new physics searches but also as a signal process to constrain the parton distribution functions (PDF) and the value of α_s itself. Since the processes that one encounters at the LHC may involve many particles in the final state — a direct consequence of the large available phase space at high energies — a precise understanding of multi-jet production is necessary.

However, the perturbative computation of the corresponding high multiplicity matrix elements is very complicated. A lot of effort has, therefore, been spent in the past to perform perturbative calculations in a fully automated way. The motivation for automation is not only to cover more processes but also to make the theoretical computations accessible to the experimental analysis in a more flexible way. For leading order (LO) tree-level calculations, this has to a large extent already been achieved. There exist various computer codes which allow a fast evaluation of the hard matrix elements and even of fully integrated cross sections [1–7]. LO computations, however, suffer from a large residual dependence on the unphysical renormalisation and factorisation scale. In addition, the QCD corrections at next-to-leading order (NLO) are quite often found to be large — of the order of 20% or even larger. Therefore, the LO result is usually sufficient only as a rough estimate. A precision prediction requires in general a full NLO computation.

At NLO, the growth in complexity for increasing multiplicity exceeds the LO case by far, the reason for this being twofold: First, the computation of the real radiation within the Catani-Seymour subtraction scheme [8, 9] requires the evaluation of a huge number of dipoles. While this problem is technically solved via automated dipole generation [10–15] and implemented with the above mentioned tree-level technology in fully automated event generators like, for example, SHERPA [10], today's computing resources still set limits on the manageable multiplicity. Second, the virtual corrections impose yet another bottleneck: On the one hand, for certain processes, the number of Feynman diagrams grows more than factorially with the number of external legs. For example, the six gluon one-loop amplitude, the dominant contribution to four-jet production, involves around 15000 diagrams (neglecting self-energy type corrections). Since every Feynman diagram represents an algebraic expression, the total number of terms can become so huge that it gets more and more difficult to simplify them even for modern computer algebra systems. On the other hand, the evaluation of a single one-loop Feynman diagram is due to the integration over the virtual degrees of freedom a non-trivial task on its own. Working in momentum space, the algebraic expression of an individual one-loop diagram is in general a product of propagators and vertices integrated over the loop momentum ℓ according to

$$\int \mathrm{d}\ell \, \frac{\mathcal{N}\left(\ell, \{K_i\}, \{M_i\}, \{m_i\}\right)}{\prod_{i=1}^{p}((\ell - K_i)^2 + m_i^2)}$$

where $\{K_i\}$ represents linear combinations of the external particles' momenta, $\{M_i\}$ the masses of external particles and $\{m_i\}$ the masses of the propagating particles in the loop. The product in the denominator stems from the propagators within the loop, while the numerator function $\mathcal{N}$, receives contributions from momentum dependent vertices and propagators. Depending on the number of propagators p and the power r of the loop momentum in the numerator, the one-loop integrals are referred to as p-point tensor integrals of rank r. In the traditional Passarino-Veltman reduction [16], tensor integrals are reduced to a set of one-point, two-point, three-point and four-point integrals with loop momentum independent numerator. These scalar master integrals are all known analytically, allowing in principle the computation of arbitrary one-loop processes. For high multiplicity, however, the algebraic reduction becomes very complicated involving large expressions multiplying the master integrals and, in addition, the reduction can lead to numerical instabilities in certain kinematic regions. Although the traditional techniques of tensor reduction could be continuously improved [17–27], this non-trivial task imposes still a big challenge for traditional Feynman diagram based computations.

Despite these severe obstacles, an impressive number of complicated QCD processes at NLO with four particles in the final state have been computed [28–48], and a hand full of $2 \rightarrow 5$ processes and even beyond that have recently appeared [49–53]. An important contribution to these high multiplicity processes came from new techniques based on unitarity which became serious alternatives to traditional Feynman diagram computations [54, 55]. The original formulation is based on the well known fact that the logarithms present in one-loop Feynman integrals develop branch cuts for distinct values of the kinematic invariants. Knowing the discontinuity across the branch cut in a kinematic channel allows a unique reconstruction of the coefficient that multiplies the corresponding integral function. On the other hand, the discontinuities across the cuts can also be computed from the Cutkosky rules [56] via a phase space integral over tree-level amplitudes by cutting the one-loop amplitude in the particular kinematic channels into two pieces. It was later realised that this formalism could be extended to multiple-cuts, often referred to as generalised unitarity, using complex momenta and spinor integration [57–60]. The remarkable point is that within unitarity methods, the so called cut-constructible part of the one-loop amplitude is effectively computed from on-shell tree-level amplitudes in four dimensions. While specific theories like for example $\mathcal{N} = 4$ Super Yang Mills are entirely cut-constructible, QCD amplitudes involve additional cut-free rational terms related to the ultraviolet behaviour of the theory. In order to compute also the remaining rational terms additional techniques have to be applied like, for example, on-shell recursion at one-loop or D-dimensional unitarity [61–67]. Another recently developed method which is directly applicable for numerical implementation is integrand reduction [68]. Within this approach, tensor reduction acts on the full one-loop integrand instead of single Feynman diagrams. The coefficients multiplying the scalar master integrals of the amplitude are thereby extracted solving a system of linear equations which is generated by evaluating the integrand for different numerical values of the loop momentum. Again, the rational terms require a separate treatment which involves among other things the derivation of effective tree-level vertices [69]. The method which unifies both the ideas from unitarity and integrand reduction for numerical application is D-dimensional generalised unitarity [70–72]. Like in the original unitarity formalism, the integrand is reconstructed from tree-level amplitudes. Yet, in order to recover the rational terms correctly, the amplitude must be computed twice in different higher integer spin dimensions and subsequently be extrapolated to the four-dimensional case.

In this work, a modified version of D-dimensional generalised unitarity in massless QCD is discussed. The method allows to compute both the cut-constructible and the rational part from tree-level amplitudes in four dimensions. The basic idea is to incorporate the additional degrees of freedom from dimensional regularisation in a uniform mass shift of the particles circulating in the loop such that the rational terms can effectively be computed in an analogous way to the cut-constructible part [62, 67]. While this works in a straightforward way in pure gauge theory where for this purpose the gluons inside the loop can simply be replaced with massive scalar particles [55, 73], scalar–quark vertices need to be defined [55, 61] in order to treat also amplitudes with external quarks. However, in this purely numerical approach, the presence of artificially massive fermion propagators leads to an extended integrand structure with additional spurious terms. We demonstrate that in the massless case, these terms occur only in the theoretically vanishing quark configurations with helicity violation along the fermion line. In our numerical approach such contributions are non-vanishing, yet, they may be set equal to zero by hand, making the algorithm effectively independent of the external helicity assignment. We mention at this point that the algorithm discussed in this work is tailored to compute primitive amplitudes — gauge invariant, purely kinematic building blocks with a fixed order of external legs which can be employed to construct the full amplitude from permutations of external particles [54, 74–76].

The great advantage of the method presented in this work with respect to D-dimensional generalised unitarity is that there is no need to perform any computation in higher dimensions allowing for efficient numerical evaluations in strictly four dimensions. Although various ingredients to the presented algorithm are known in principle, there does not exist a concise write-up which explains how the individual constituents can be combined for a fully numerical and automated computation of massless one-loop QCD amplitudes.

Another important intention of this project has been to make a numerical implementation of the outlined techniques publicly available. For this purpose, we published in a first step the program NGLUON which is able to evaluate pure gluonic primitive amplitudes at one-loop order with arbitrarily many external gluons and any helicity configuration [77]. NGLUON was extended to deal also with an arbitrary number of external quark–anti-quark pairs [78] and, finally, to compute colour and helicity summed interferences between Born and virtual corrections expressed in terms of primitives [79]. All results are collected in the publicly available program NJET which provides besides the mere primitive amplitudes all necessary colour and helicity summed virtual corrections contributing to two-jet, three-jet, four-jet and even five-jet production [80]. Several collaborations have also presented attempts for automation of one-loop virtual corrections [81–90]. So far only about half of them are publicly available codes. This is, however, already much more than when this project started. We believe, therefore, that the public release of NGLUON and NJET has to a certain extent triggered other collaborations to make their programs publicly available.

As the final result, NJET was successfully applied to compute four-jet production at NLO for the LHC [91]. While the first full NLO computations of two-jet production at hadron colliders date back roughly 20 years [92, 93], it took another decade for the full completion of three-jet production in hadronic collisions [94–97]. The enormous complexity of calculating four-jet production at NLO is the main reason why only recently, first NLO results of this challenging process for the LHC at a centre of mass energy $\sqrt{s} = 7$ TeV were available [98]. We could confirm these results — a non-trivial check taking the difficulty of the computation into account — and present new results at $\sqrt{s} = 8$ TeV with substantial new insight with respect to the size of the corrections.

The outline of the present work is the following: In chapter 1, we review known colour decompositions that express both tree-level and one-loop QCD amplitudes in terms of primitive amplitudes. In chapter 2, we describe the above mentioned algorithm based on unitarity and integrand reduction for the numerical evaluation of one-loop primitive amplitudes in massless QCD. In chapter 3, the validation, the numerical accuracy as well as the runtime performance of NGLUON and NJET are discussed. Finally, in chapter 4, results for four-jet production at NLO for the LHC at a centre of mass energy of $\sqrt{s} = 8$ TeV obtained from NJET are presented.

1. Colour ordering and primitive amplitudes

In this chapter, known colour decompositions of both tree-level and one-loop QCD amplitudes are reviewed. The intention is to set up the necessary preliminaries for the computation of one-loop amplitudes with unitarity methods and integrand reduction techniques, subject of chapter 2. Colour decompositions express the amplitude as sums of independent colour structures multiplied with purely kinematic factors, the so called partial amplitudes. In general, partial amplitudes can be further decomposed into primitive amplitudes, gauge invariant building blocks with a fixed order of external legs and a much simpler kinematic structure. At tree-level, both the colour decompositions and well established methods for the evaluation of primitive amplitudes are reviewed, in section 1.1 for the pure gluonic case and in section 1.2 for the mixed quark-gluon case. The decompositions of one-loop amplitudes in terms of primitives is discussed in section 1.3. Finally, in section 1.4, the colour summation in terms of primitives for the squared Born and for the interference of the Born with the virtual corrections — quantities which enter the computation of cross sections at next-to-leading order — is discussed.

1.1. Tree-level amplitudes in pure gauge theory

Pure gluon amplitudes at tree-level can be decomposed into sums of independent colour structures multiplied with a unique purely kinematic factor known as the *partial amplitude* or *colour ordered amplitude* [99]:

$$
\mathcal{A}_n^{\text{tree}}(\{p_1, h_1, a_1\}, \ldots, \{p_n, h_n, a_n\}) =
$$
$$
\sum_{\sigma \in S_n / Z_n} \text{Tr}(T^{a_{\sigma(1)}} T^{a_{\sigma(2)}} \ldots T^{a_{\sigma(n)}}) A_n^{\text{tree}}(\sigma(1), \sigma(2), \ldots, \sigma(n)). \quad (1.1)
$$

T^a with $a = 1, \ldots, N_c^2 - 1$ are the generators of the $\mathfrak{su}(N_c)$ Lie algebra in the fundamental representation defined via the commutation relation

$$
[T^{a_1}, T^{a_2}] = i\sqrt{2} f^{a_1 a_2 x} T^x \quad (1.2)
$$

and f^{abc} being the structure constants. In the following, we assume always a summation over repeated colour indices. Note that the structure constants form itself a real representation of the $\mathfrak{su}(N_c)$ Lie algebra, the adjoint representation. The fundamental generator matrices are normalised according to

$$\text{Tr}(T^{a_1}T^{a_2}) = \delta^{a_1 a_2}. \tag{1.3}$$

The sum in Eq. (1.1) denotes the non-cyclic permutations S_n/Z_n of the external particle labels. The kinematic part $A_n^{\text{tree}}(\sigma(1), \sigma(2), \ldots, \sigma(n))$ depends only on the external momenta p_i and helicities h_i and on the permutation σ of the external particles. The integer arguments $\sigma(i)$ are a shorthand notation to represent both the momentum and the helicity of the ith particle at the same time, i.e.

$$A_n^{\text{tree}}(1, 2, \ldots, n) \equiv A_n^{\text{tree}}(\{p_1, h_1\}, \{p_2, h_2\}, \ldots, \{p_n, h_n\}). \tag{1.4}$$

Any partial amplitude is gauge invariant, invariant under cyclic permutations of the external legs and obeys the reflection identity $A(1, \ldots, n) = (-1)^n A(n, \ldots, 1)$ [99]. The aim of this section is to review some of the methods to compute colour ordered gluon amplitudes based on arguments presented in Refs. [99–101].

We employ the QCD Feynman rules in Lorentz Feynman gauge as stated in Ref. [99]. The gluon propagator P_g, the three-gluon vertex V_{3g} with all momenta outgoing and the four-gluon vertex V_{4g} thus read

$$P_g = \frac{-i}{p^2 + i\varepsilon} g_{\mu_1 \mu_2} \delta^{a_1 a_2}, \tag{1.5}$$

$$V_{3g} = -g_s f^{a_1 a_2 a_3} \left\{ g_{\mu_1 \mu_2}(p_1 - p_2)_{\mu_3} + g_{\mu_2 \mu_3}(p_2 - p_3)_{\mu_1} + g_{\mu_3 \mu_1}(p_3 - p_1)_{\mu_2} \right\}, \tag{1.6}$$

$$V_{4g} = -ig_s^2 \sum_{C(1,2,3)} f^{a_1 a_2 x} f^{x a_3 a_4} \left\{ g_{\mu_1 \mu_3} g_{\mu_2 \mu_4} - g_{\mu_1 \mu_4} g_{\mu_2 \mu_3} \right\}. \tag{1.7}$$

with the Feynman ε-prescription in the propagator.[1] g_s is the non-abelian gauge coupling and $C(1, 2, 3)$ denotes the cyclic permutations of the index list $[1, 2, 3]$. The sign convention for the metric tensor is $g_{\mu\nu} = \text{diag}(+1, -1, -1, -1)$. Combining Eq. (1.2) and (1.3), it follows that any adjoint colour matrix f^{abc} can be expressed in terms of traces of fundamental colour matrices

$$f^{a_1 a_2 a_3} = -\frac{i}{\sqrt{2}} \text{Tr}([T^{a_1}, T^{a_2}]T^{a_3}). \tag{1.8}$$

One can thus rewrite the gluon vertices in terms of fundamental colour traces leading to

$$V_{3g} = g_s \sum_{P(1,2)} \text{Tr}(T^{a_1} T^{a_2} T^{a_3})$$

$$\frac{i}{\sqrt{2}} \left\{ g_{\mu_1 \mu_2}(p_1 - p_2)_{\mu_3} + g_{\mu_2 \mu_3}(p_2 - p_3)_{\mu_1} + g_{\mu_3 \mu_1}(p_3 - p_1)_{\mu_2} \right\}, \tag{1.9}$$

---

[1]We will suppress in practical calculations $+i\varepsilon$ in the propagators as long as no ambiguities arise.

$$V_{4g} = g_s^2 \sum_{C(1,2,3)} \text{Tr}([T^{a_1}, T^{a_2}][T^{a_3}, T^{a_4}]) \frac{i}{2} \Big\{ g_{\mu_1\mu_3} g_{\mu_2\mu_4} - g_{\mu_1\mu_4} g_{\mu_2\mu_3} \Big\}$$

$$= g_s^2 \sum_{P(1,2,3)} \text{Tr}(T^{a_1} T^{a_2} T^{a_3} T^{a_4})$$

$$\frac{i}{2} \Big\{ 2g_{\mu_1\mu_3} g_{\mu_2\mu_4} - g_{\mu_1\mu_2} g_{\mu_3\mu_4} - g_{\mu_1\mu_4} g_{\mu_3\mu_2} \Big\} \tag{1.10}$$

where $P(1, \ldots, m)$ denotes the $m!$ permutations of m indices. The Lorentz part defines the *colour ordered Feynman rules*[2] [74, 101]:

$$P_g^{\text{co}} = \frac{-i}{p^2 + i\varepsilon} g_{\mu_1\mu_2}, \tag{1.11}$$

$$V_{3g}^{\text{co}} = g_s \frac{i}{\sqrt{2}} \Big\{ g_{\mu_1\mu_2}(p_1 - p_2)_{\mu_3} + g_{\mu_2\mu_3}(p_2 - p_3)_{\mu_1} + g_{\mu_3\mu_1}(p_3 - p_1)_{\mu_2} \Big\}, \tag{1.12}$$

$$V_{4g}^{\text{co}} = g_s^2 \frac{i}{2} \Big\{ 2g_{\mu_1\mu_3} g_{\mu_2\mu_4} - g_{\mu_1\mu_2} g_{\mu_3\mu_4} - g_{\mu_1\mu_4} g_{\mu_3\mu_2} \Big\}. \tag{1.13}$$

Since the colour is stripped off, the vertices are not Bose symmetric. Colour information is still implicitly contained in the vertices by means of the order of legs attached to the vertex: To every ordering belongs a unique colour trace with equal ordering of generator matrices. This property translates also to the diagrammatic level: A tree-level Feynman diagram matching a fixed order of external legs, drawn with the above colour ordered vertices belongs uniquely to that partial amplitude whose order of generators in the colour trace $\text{Tr}(T^{a_1} T^{a_2} \ldots T^{a_n})$ is the same as the order of external legs. In order to see why this is true, one can restore at every vertex within the diagram the colour traces, followed by a summation over the internal gluon indices using the Fierz identity

$$T^x_{i_1 \bar{i}_2} T^x_{i_3 \bar{i}_4} = \delta_{i_1 \bar{i}_4} \delta_{i_3 \bar{i}_2} - \frac{1}{N_c} \delta_{i_1 \bar{i}_2} \delta_{i_3 \bar{i}_4} \tag{1.14}$$

and neglecting all subleading colour terms since they do not contribute to the amplitude.[3] Hence, the sum of all *colour ordered Feynman diagrams* whose external legs match the order of generator matrices within the trace $\text{Tr}(T^{a_1} T^{a_2} \ldots T^{a_n})$ is precisely the Lorentz part $A_n^{\text{tree}}(1, 2, \ldots, n)$ in Eq. (1.1). A representation of polarisation vectors of the external gluons to evaluate the colour ordered diagrams is given in appendix A.

A very efficient and elegant method to sum up the colour ordered diagrams is the famous Berends-Giele recursion relation, the pioneering work of colour ordered gluon computations [99]. The basic idea is the following: Removing from an n-point par-

---

[2]Note that we included also the factors of $\frac{1}{\sqrt{2}}$ from the colour part as in Refs. [74, 101].

[3]An algorithmic way to see the absence of subleading colour contributions in Eq. (1.1) is the following: An arbitrary Feynman diagram contributing to n-gluon scattering involves products of $(n-2)$ adjoint colour matrices multiplying a kinematic part. Expressing one arbitrary matrix f^{abc} via fundamental colour traces with help of Eq. (1.8), followed by the repeated application of Eq. (1.2), the colour factor becomes a particular combination of commutators of fundamental generators within the trace. Expanding the commutators of all diagrams and collecting equal colour structures exploiting the cyclic symmetry of the trace, one arrives at the decomposition in Eq. (1.1).

tial gluon amplitude $A_n(1, \ldots, n)$ the nth polarisation vector and allowing the four-momentum to be off the mass-shell, one formally ends up with a vector current $J_\mu(1, \ldots, n-1)$ which corresponds to the production of $(n-1)$ on-shell gluons (all momenta are always taken to be outgoing). Following the off-shell leg into the diagrams, one arrives either at a three-gluon vertex or at a four-gluon vertex. At the three-gluon vertex, the current is split into all combinations of two smaller subcurrents $J_\mu(1, \ldots, i)$ and $J_\mu(i+1, \ldots, n-1)$ that have the same ordering as the $(n-1)$-point current $J_\mu(1, \ldots, n-1)$. Analogously at the four-gluon vertex, the current is split into all combinations of three smaller subcurrents $J_\mu(1, \ldots, i)$, $J_\mu(i+1, \ldots, j)$ and $J_\mu(j+1, \ldots, n-1)$ that respect the ordering of the $(n-1)$-point current. In the same manner, every subcurrent is recursively split up again as many times until one hits a one-point current which is then identified with the gluon polarisation vector of momentum p_i and helicity h_i

$$J_\mu(i) = \epsilon_\mu^{(h_i)}(p_i). \tag{1.15}$$

Defining the *region momenta* $P_{i,j}$ as the momentum sum

$$P_{i,j} = \sum_{k=i}^{j} p_k \qquad \text{for} \quad j \geq i, \tag{1.16}$$

the Berends-Giele recursion in Lorentz Feynman gauge reads [99]

$$J^\mu(1, \ldots, n) = \frac{-i}{P_{i,n}^2} \left[\sum_{i=1}^{n-1} \widehat{V}_{3g}^{\alpha_1 \alpha_2; \mu}(P_{1,i}, P_{i+1,n}) \, J_{\alpha_1}(1, \ldots, i) \, J_{\alpha_2}(i+1, \ldots, n) \right. \tag{1.17}$$

$$\left. + \sum_{i=1}^{n-2} \sum_{j=i+1}^{n-1} \widehat{V}_{4g}^{\alpha_1 \alpha_2 \alpha_3 \mu} \, J_{\alpha_1}(1, \ldots, i) J_{\alpha_2}(i+1, \ldots, j) J_{\alpha_3}(j+1, \ldots, n) \right].$$

The vertices $\widehat{V}_{3g}$ and $\widehat{V}_{4g}$ follow from Eqs. (1.12) and (1.13) exploiting current conservation $P_{i,j}^\mu J_\mu(i, \ldots, j) = 0$ and momentum conservation in the three-point vertex $P_1 + P_2 + P_3 = 0$:

$$\widehat{V}_{3g}^{\alpha_1 \alpha_2; \mu}(P_1, P_2) = g_s \frac{i}{\sqrt{2}} \left\{ g^{\alpha_1 \alpha_2}(P_1 - P_2)^\mu + 2g^{\alpha_2 \mu} P_2^{\alpha_1} - 2g^{\mu \alpha_1} P_1^{\alpha_2} \right\}, \tag{1.18}$$

$$\widehat{V}_{4g}^{\alpha_1 \alpha_2 \alpha_3; \mu} = g_s^2 \frac{i}{2} \left\{ 2g^{\alpha_1 \alpha_3} g^{\alpha_2 \mu} - g^{\alpha_1 \alpha_2} g^{\alpha_3 \mu} - g^{\alpha_1 \mu} g^{\alpha_3 \alpha_2} \right\}. \tag{1.19}$$

An n-point colour-ordered tree-level gluon amplitude is now simply computed as the on-shell contraction of an $(n-1)$-point gluon current with the nth polarisation vector

$$A_n^{\text{tree}}(1, \ldots, n) = i P_{1,n-1}^2 \, J^\mu(1, \ldots, n-1) \, J_\mu(n) \Big|_{P_{1,n}=0}. \tag{1.20}$$

Note that due to the on-shell contraction, the current $J^\mu(1, \ldots, n-1)$ must be multiplied with the inverse on-shell propagator $i P_{1,n-1}^2 = i p_n^2$. The algorithmic realisation of the Berends-Giele recursion as it is implemented in the NGLUON and NJET library is described in appendix B.

In recent years, new on-shell methods have revealed new techniques which do not rely on the explicit computation of Feynman diagrams any more. With the meanwhile well established BCFW recursion, on-shell amplitudes of distinct multiplicity are constructed recursively from lower point on-shell amplitudes [102]. Reviews about the development in this field can be found, for example, in Refs. [103, 104]. Throughout this work, however, we use exclusively Berends-Giele recursion for the numerical evaluation of colour ordered tree-level amplitudes.

We conclude this section with a remark on the number of required permutations of external gluons in the colour ordered amplitudes to compute the full colour amplitude. From the trace basis in Eq. (1.1) one expects $(n-1)!$ independent amplitude evaluations which are reduced with help of the reflection identity by another factor of two (for details see Ref. [99]). Kleiss and Kuijf established additional linear relations between colour ordered amplitudes which reduce the number of necessary amplitudes down to $(n-2)!$ [105]. The colour decomposition given in Ref. [106] finally expresses the full colour amplitude in terms of $(n-2)!$ colour ordered amplitudes, though, it uses strings of adjoint colour matrices as basis vectors instead of traces of fundamental generators:

$$\mathcal{A}_n^{\text{tree}} = \sum_{\sigma \in P(2,\dots,n-1)} (F^{a_{\sigma(2)}} \dots F^{a_{\sigma(n-1)}})_{a_1 a_n} A_n^{\text{tree}}(1, \sigma(2), \dots, \sigma(n-1), n) \qquad (1.21)$$

with the definition

$$F_{bc}^a \equiv -i\sqrt{2}f^{abc} \qquad \text{and} \qquad \text{Tr}(F^a F^b) = 2N_c \delta^{ab}. \qquad (1.22)$$

We mention in this context also the recently found BCJ-relations [107] which in principle allow a reduction down to $(n-3)!$ partial amplitudes. From a practical point of view, the implementation of these relations is not straightforward.

1.2. Quark–gluon tree-level amplitudes

We employ again the conventions for the QCD Feynman rules from Ref. [99]. The quark propagator and the quark–gluon vertex read

$$P_q = \frac{i(\not{p} + m)}{p^2 - m^2 + i\varepsilon} \delta_{i\bar{j}}, \qquad (1.23)$$

$$V_{qg\bar{q}} = i\, g_s\, T_{ij}^a\, \gamma^\mu \qquad (1.24)$$

where $\not{p} \equiv p_\mu \gamma^\mu$ and m is the particle mass. A representation of the gamma matrices γ^μ and of the corresponding spinors are given in appendix A. A frequently used colour decomposition of QCD tree-level amplitudes with arbitrarily many quarks and gluons is discussed in Ref. [100]: A distinct colour structure that multiplies the colour stripped Lorentz part of a quark–gluon amplitude with m quark lines and n emitted gluons is a product of m strings of generator matrices where every string is limited by two open fundamental indices, one belonging to a quark and the other one to an anti-quark:

$$\hat{c}(\{n_j\},\{\bar{\alpha}\}) = \frac{(-1)^p}{N_c^p}(T^{a_1}\ldots T^{a_{n_1}})_{i_1\bar{\alpha}_1}(T^{a_{n_1+1}}\ldots T^{a_{n_2}})_{i_2\bar{\alpha}_2}\ldots(T^{a_{n_{m-1}+1}}\ldots T^{a_n})_{i_m\bar{\alpha}_m}.$$
(1.25)

The indices $n_1,\ldots,n_{m-1}$ with $1 \leq n_j \leq n$ denote "... an arbitrary partition of an arbitrary permutation of the n gluon indices." [100] The generator string shrinks to a Kronecker delta in case that no external gluon is emitted from a distinct colour flow line connecting two fundamental indices. The factor $(-1)^p/N_c^p$ counts the number of gluon propagators which do not mediate colour flow between two quark lines (for this reason they are also called QCD photons or $U(1)$-gluons). The range of p is therefore $0 \leq p \leq m-1$. The full amplitude is then the sum of all colour structures $\hat{c}_i$ multiplied by the purely kinematic partial amplitudes $A_{i,\{q,g\}}^{\text{partial}}(1,\ldots,n)$

$$\mathcal{A} = \sum_i \hat{c}_i \, A_{i,\{q,g\}}^{\text{partial}}(1,\ldots,n).$$
(1.26)

For more details we refer to Ref. [100]. In the pure gluonic case, every colour stripped partial amplitude $A_{\{g\}}^{\text{tree}}(1,\ldots,n)$ is automatically colour ordered which means that to every colour structure, a unique amplitude with a fixed order of external legs belongs being computable with the colour ordered Feynman rules in Eqs. (1.11) – (1.13). Although this statement can be extended to amplitudes with a single quark line plus arbitrarily many gluons [99], the uniqueness is in general lost for amplitudes with more than one quark line. This is illustrated at the simplest possible example, a four quark amplitude with two different quark flavours consisting of a single Feynman diagram:

$$\mathcal{A}(q,\bar{q},Q,\bar{Q}) = \begin{matrix} \bar{i}_2 \\ \\ i_1 \end{matrix}\!\!\!\!\!\!\!\!\!\!\!\!\!\!\!\!\!\!\begin{matrix} i_3 \\ \text{OOO} \\ \bar{i}_4 \end{matrix} = T_{i_1\bar{i}_2}^x T_{i_3\bar{i}_4}^x A(q,\bar{q},Q,\bar{Q})$$

$$= (\delta_{i_1\bar{i}_4}\delta_{i_3\bar{i}_2} - \frac{1}{N_c}\delta_{i_1\bar{i}_2}\delta_{i_3\bar{i}_4})\, A(q,\bar{q},Q,\bar{Q}).$$

While the first colour structure $\delta_{i_1\bar{i}_4}\delta_{i_3\bar{i}_2}$ mediates colour flow between the two quark lines, the $1/N_c$ suppressed term — the subleading colour contribution — does not. Yet, the Lorentz structure $A(q,\bar{q},Q,\bar{Q})$ is the same for both independent colour structures. Hence, although the colour matrix T_{ij}^a of the quark–gluon vertex can be stripped off, the remaining vertex is not colour ordered, because, in contrast to the pure gluonic case, we are not allowed to discard all subleading colour contributions when restoring the colour information and applying the Fierz identity from Eq. (1.14). One approach to deal with the subleading colour contributions that still allows a separate treatment of the colour part and the kinematic part is to work with QCD in the colour flow representation [108]. The colour ordered recursion relations as we encountered them in the gluonic case, however, require some modification [109]. We will therefore not follow this approach any further. More information about QCD in the colour flow representation can be found, for example, in Ref. [110].

One can achieve colour ordering at the level of colour stripped amplitudes also for arbitrary quark–gluon amplitudes if one works with quarks in the adjoint representation [74]. The corresponding adjoint quark–gluon vertex reads

$$V_{qg\bar{q}}^{\text{adj}} = g_s f^{a_1 a_2 a_3} \gamma^\mu. \tag{1.27}$$

Such a vertex is, for example, present in $\mathcal{N} = 4$ Super Yang Mills (SYM) theory whose lagrangian was first described in Ref. [111]. Like in the pure gluonic case, one can re-express the adjoint colour matrices with traces of fundamental ones resulting effectively into two *colour ordered quark–gluon vertices* which are anti-symmetric with respect to flipping legs [74, 101]:

$$\vcenter{\hbox{\includegraphics{diagram1}}} = + g_s \frac{i}{\sqrt{2}} \gamma^\mu \qquad \vcenter{\hbox{\includegraphics{diagram2}}} = - g_s \frac{i}{\sqrt{2}} \gamma^\mu. \tag{1.28}$$

We refer in the following to the vertex with positive sign as *even ordered* and to the one with negative sign as *odd ordered*. In analogy to the pure gluonic case, a colour ordered quark–gluon amplitude is the sum of all colour ordered Feynman diagrams drawn with the above two quark–gluon vertices and the pure gluonic ones in Eqs. (1.12) and (1.13), all being consistent with a fixed order of external legs and flavours. Here, "flavour" characterises both quarks and gluons. In the following we will refer to these colour ordered tree-level amplitudes as *tree-level primitive amplitudes*. In $\mathcal{N} = 4$ SYM amplitudes, besides gluons and adjoint fermions (gluinos), there exist also scalars which couple both to fermions, gluons and to themselves. It can be shown that the primitive amplitudes for massless quarks are implicitly present in $\mathcal{N} = 4$ SYM theory as special linear combinations of gluon-gluino tree amplitudes [112], i.e. that within such linear combinations all contributions with internal scalars are missing.

One of the great advantages of working with quarks in the adjoint representation is that due to the colour ordering of the vertices, one can compute primitive amplitudes with the same recursive off-shell techniques as described in the previous section. The basic difference is that every off-shell current carries now also an additional flavour index which can either be a quark, an anti-quark or a gluon. The algorithmic realisation is then a priori a matter of bookkeeping which makes sure that only currents of equal flavour will be connected. In appendix B.2, we present a quark-index-system which allows the computation of primitive amplitudes via off-shell recursion with an arbitrary number of quark pairs. A detailed study which compares amplitudes computed via off-shell recursion with analytical formulae obtained from on-shell recursion has recently been done in Ref. [113].

The remaining non-trivial aspect is to relate the primitive amplitudes to the partial amplitudes in QCD. The general idea how to do that has been described in Ref. [75] which has subsequently been applied in a more automated way to obtain the colour decomposition for QCD amplitudes with up to seven external partons both at tree-level and at one-loop order [76]. The method employed in NJET to obtain the partial amplitudes of any quark–gluon amplitude is very similar and shall be described in the remainder of this section.

The ith partial amplitude $A_{i,\{q,g\}}^{\text{partial}}$ in Eq. (1.26) receives colour stripped kinematical contributions K_j from various Feynman diagrams. We may therefore write

$$A_{i,\{q,g\}}^{\text{partial}} = \sum_{j=1}^{N_K} S_{ij} K_j \tag{1.29}$$

where N_K is the total number of colour stripped kinematic contributions. The entries of the matrix S_{ij} are either 0 or 1, depending on which particular K_j contributes. On the other hand, since the additional odd colour stripped adjoint quark–gluon vertex differs from the QCD vertex only by an additional minus sign, the primitive amplitudes are linear combinations of the same K_j present in the QCD partial amplitudes, however at any time weighted with a factor $(-1)^q$ where q denotes the number of odd quark–gluon vertices. Hence, any primitive amplitude $\mathcal{P}_l$ can be expressed as

$$\mathcal{P}_l = \sum_{m=1}^{N_K} M_{lm} K_m \tag{1.30}$$

where the entries of the matrix M are either 0 or ±1. This system of equations can be solved for K_m

$$K_m = \sum_{l=1}^{N_P} B_{ml} \mathcal{P}_l \tag{1.31}$$

with N_P being the total number of primitive amplitudes. Plugging the solution into Eq. (1.29), a partial amplitude in terms of primitive amplitudes thus reads

$$A_{i,\{q,g\}}^{\text{partial}} = \sum_{j=1}^{N_K} \sum_{k=1}^{N_P} S_{ij} B_{jk} \mathcal{P}_k \ . \tag{1.32}$$

Note that the system of equations in (1.30) is in general overconstrained if one considers indeed all possible primitive amplitudes $\mathcal{P}_l$. The matrix M contains thus a non-trivial null-space. Putting M into reduced row echelon form, one can always achieve a decomposition of the full colour QCD amplitude in terms of a minimal number of independent primitive amplitudes $\hat{N}_P$ which is in general smaller than the total number of primitives N_P. As a byproduct, one gets $(N_P - \hat{N}_P)$ relations between dependent primitive amplitudes, a direct consequence of the non-trivial null space of the matrix

M. For this reason, the decomposition is in general not unique.

The kinematic contributions K_j need not be evaluated to get the colour decomposition, they should rather be considered as a tag belonging to a certain diagram, or to a certain group of diagrams. In practice, one may first generate all Feynman diagrams of an amplitude (for example with QGRAF [114]), collect equal colour structures and determine in this way the matrix S. In order to determine the matrix M, one maps the topologies of the colour stripped parts to the colour ordered diagrams that contribute to all possible primitive amplitudes.

To be explicit, we give as an example the decomposition of the full colour QCD amplitude $\mathcal{A}(q, \bar{q}, Q, \bar{Q}, g)$ with two unequal flavour quark lines plus one gluon in terms of primitives:

$$\mathcal{A}(q, \bar{q}, Q, \bar{Q}, g) = (T^{a_5})_{i_1 \bar{i}_4} \delta_{i_3 \bar{i}_2} \mathcal{P}_{[12345]} + \delta_{i_1 \bar{i}_4} (T^{a_5})_{i_3 \bar{i}_2} \mathcal{P}_{[12534]} + \frac{1}{N_c} \delta_{i_1 \bar{i}_2} (T^{a_5})_{i_3 \bar{i}_4} \mathcal{P}_{[12354]}$$

$$- \frac{1}{N_c} (T^{a_5})_{i_1 \bar{i}_2} \delta_{i_3 \bar{i}_4} \left(\mathcal{P}_{[12345]} + \mathcal{P}_{[12354]} + \mathcal{P}_{[12534]} \right). \tag{1.33}$$

The notation $\mathcal{P}_{[\sigma_1 \dots \sigma_5]}$ means a primitive amplitude with the order of external particles being a distinct permutation of the list $[12345]$ where we used the shorthand notation $1 = q, 2 = \bar{q}, 3 = Q, 4 = \bar{Q}$ and $5 = g$. There are four independent colour structures and three independent primitive amplitudes. As explained above, the decomposition is not unique: For example, the partial amplitude belonging to the colour structure $N_c^{-1} (T^{a_5})_{i_1 \bar{i}_2} \delta_{i_3 \bar{i}_4}$ could also be expressed in terms of another primitive $\mathcal{P}_{[15234]}$. Yet, with the relation $\mathcal{P}_{[12354]} = -(\mathcal{P}_{[15234]} + \mathcal{P}_{[12534]} + \mathcal{P}_{[12345]})$ the minimal number of primitives is still three.

1.3. Colour decomposition and primitive amplitudes at one-loop order

In analogy to the tree-level techniques described in sections 1.1 and 1.2, we first discuss the one-loop colour decomposition of pure gluon amplitudes and subsequently the more difficult mixed quark–gluon case.

In Ref. [115], a colour decomposition of one-loop amplitudes in pure gauge theory in terms of single and double trace objects is given. Including also the n_f-dependent terms with a closed quark loop [116], this reads

$$\mathcal{A}_n^{\text{1-loop}} = g_s^n \left[\sum_{j=1}^{\lfloor n/2 \rfloor + 1} \sum_{\sigma \in S_n / S_{n;j}} \text{Gr}_{n;j}(\sigma_1, \dots, \sigma_n) A_{n;j}^{[1]}(\sigma_1, \dots, \sigma_n) \right.$$

$$\left. + \frac{n_f}{N_c} \sum_{\sigma \in S_n / S_{n;1}} \text{Gr}_{n;1}(\sigma_1, \dots, \sigma_n) A_{n;1}^{[1/2]}(\sigma_1, \dots, \sigma_n) \right] \tag{1.34}$$

where

$$\mathrm{Gr}_{n;1}(1, 2, \ldots, n) = \mathrm{Tr}(1)\,\mathrm{Tr}(T^{a_1}T^{a_2}\ldots T^{a_n}) = N_c\,\mathrm{Tr}(T^{a_1}T^{a_2}\ldots T^{a_n}),$$
$$\mathrm{Gr}_{n;j}(1, \ldots, j-1, j, \ldots, n) = \mathrm{Tr}(T^{a_1}\ldots T^{j-1})\,\mathrm{Tr}(T^{a_j}\ldots T^{a_n}).$$

and $\lfloor x \rfloor$ is the largest integer less than or equal to x. $S_n/S_{n;j}$ are those permutations which leave the trace structures $\mathrm{Gr}_{n;j}$ invariant. $A_{n;j}^{[1]}$ are the corresponding one-loop partial amplitudes for each independent colour structure. The superscript [1] indicates that there are only spin one particles circulating in the loop. The single trace contributions $\mathrm{Gr}_{n;1}$ multiplying $A_{n;1}^{[1]}$ are enhanced by a factor of N_c, the leading colour contributions. They are colour ordered and can therefore be computed with the colour ordered Feynman rules in Eqs. (1.12) and (1.13): Like in the tree-level case, they are the sum of all colour ordered one-loop Feynman diagrams which match a fixed order of the external particles. These objects are in analogy to the tree-level case called *one-loop primitive amplitudes* and are gauge invariant [115]. It can be shown that the subleading colour partial amplitudes for the case $j > 1$ can all be obtained as special linear combinations of the leading colour partial amplitudes [54].

The n_f dependent contributions with a closed quark loop are much simpler because there are only single trace structures present, i.e. all partial amplitudes $A_{n;j}^{[1/2]}$ for $j > 1$ vanish where the superscript [1/2] denotes the pure fermionic loop content. With respect to the pure gluonic case, the remaining single trace contributions are suppressed by a factor of $1/N_c$. The much simpler colour structure can be understood from a diagramwise colour stripping. If we take an arbitrary Feynman diagram and remove all external trees which are attached to the quark loop we get a single trace structure of internal adjoint indices $\mathrm{Tr}(T^{x_1}\ldots T^{x_m})$ with $m \leq n$. Starting from this trace, one can replace step by step all adjoint colour matrices from the attached trees by means of the commutator relation (1.2), precisely as in the tree-level case. This leads to the same colour structure like in tree-level gluon amplitudes: single trace objects. The crucial point is again that the partial amplitudes $A_n^{[1/2]}$ — the primitive amplitudes with fermion loop — are colour ordered and can be computed as the sum of all colour ordered one-loop Feynman diagrams matching an external order of particles, and where within the quark loop only the even ordered quark-gluon vertex from Eq. (1.28) is used.

For completeness, we give also the full one-loop gluon amplitude with the colour decomposition of Ref. [106]:

$$\mathcal{A}_n^{\text{1-loop}} = g_s^n \sum_{S_{n-1}/\mathcal{R}} \left[\mathrm{Tr}(F^{a_{\sigma_1}}\ldots F^{a_{\sigma_n}})A_{n;1}^{[1]}(\sigma_1, \ldots, \sigma_n) \right.$$

$$\left. + 2n_f \mathrm{Tr}(T^{a_{\sigma_1}}\ldots T^{a_{\sigma_n}})A_{n;1}^{[1/2]}(\sigma_1, \ldots, \sigma_n) \right]. \qquad (1.35)$$

The F matrices are defined in section 1.1 and the reflection $\mathcal{R}$ of a permutation is defined as $\mathcal{R}([1, \ldots, n]) = [n, \ldots, 1]$. The reason for excluding the reflections is the anal-

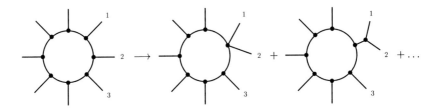

Figure 1.1.: Pinching and pulling external legs from the parent diagram to generate all colour ordered diagrams. The lines stand for both quarks and gluons. Created diagrams with vertices that do not exist are discarded, like for example when pinching together two differently flavoured quark lines or a quark–anti-quark pair attached via a four-point vertex to the loop.

ogous symmetry present also at tree-level $A_{n;1}^{[J]}(1,\ldots,n) = (-1)^n A_{n;1}^{[J]}(n,\ldots,1)$ with $[J] \in \{[1/2],[1]\}$ [106, 115]. Although this decomposition is maybe less intuitive, it is more efficient since it involves only single trace objects. We emphasise that $A_{n;1}^{[J]}$ are the same primitive amplitudes for both colour decompositions.

Comparing the one-loop primitive amplitude with their tree-level counterpart from the previous sections, its unique determination requires besides the external order of particles also a specification of the exact loop content. An arbitrary one-loop primitive amplitude (now also including external quarks) is therefore conveniently defined with help of a so called *parent diagram* [74]. A parent diagram belonging to an n-point primitive amplitude is an abstract one-loop diagram with n external ordered legs and n loop propagators, i.e. there are n external legs attached directly to the loop exclusively via three-point vertices. Its special property is that all colour ordered Feynman diagrams of a distinct one-loop primitive amplitude can be obtained from the parent diagram by pinching and pulling external legs without changing their total order. This technique is schematically illustrated in Fig. 1.1: Assuming leg 1 and 2 and the propagator in between are gluons, pinching the two legs results in a diagram with a four-gluon vertex attached to the loop, pulling this vertex away from the loop results in an external tree attached to the loop. If leg 1 and 2 form a quark–anti-quark pair connected by the single quark propagator between the two external legs, the pinching operation results formally in a diagram with a non-existing 4-point interaction which will be discarded, however, pulling the quark pair further away from the loop gives again a valid colour ordered diagram with an external quark–anti-quark pair attached to the loop via a gluon propagator. In case that leg 1 and 2 form still a quark–anti-quark pair, yet, with the propagator between them being a gluon, the quark line must be traced through all remaining other propagators of the parent diagram with the order of the external particles still being the same. This different routeing of the fermion flow contributes to a different QCD colour structure. The parent diagram forms thus a different primitive amplitude with different colour ordered diagrams.

In Ref. [74], the two different routeings of the fermion flow from one external quark

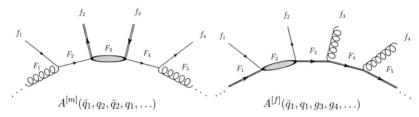

Figure 1.2.: The two possible assignments of the parent diagram given the flavour list of external particles $\mathcal{F}_{\text{ext}} = [f_1, \ldots, f_n]$ and the flavour of the first propagator in the loop F_1 as a seed to determine all subsequent ones: A gluon leads to the mixed quark–gluon case, a quark with a loop flavour different from all external quarks to the closed quark loop case. The grey blobs represent dummy propagators in case the corresponding three-point vertex does not exist.

pair (the remaining legs being gluons exclusively) have been classified as left handed primitives A_n^L if the fermion flow turns left relative to the loop and as right handed primitives A_n^R if the fermion flow turns right relative to the loop. It is further shown that these two subclasses of primitive amplitudes are separately gauge invariant which holds also in the case of a closed quark loop. Exploiting the anti-symmetry of the colour ordered Feynman rules, it is straightforward to show the mirror symmetry

$$A_n^{R,[J]}(\bar{q}, 3, 4, \ldots, q, \ldots, n-1, n) = (-1)^n A_n^{L,[J]}(\bar{q}, n, n-1, \ldots, q, \ldots, 4, 3) \qquad (1.36)$$

where $[J]$ denotes an arbitrary loop content, either mixed quark–gluon case or a closed quark loop [74]. In case of multiple fermion lines one can assign every quark pair its own handedness, as described in Ref. [76]. The same arguments as for the single quark line case can be applied to show that a primitive amplitude $A_n^{\ldots L_i R_j \ldots}$ is gauge invariant for any valid assignments of fermion routeings, and an analogous mirror symmetry as in the single quark pair case holds [76]. The complexity of the left-right assignments of any single quark line can be circumvented respecting the following two conventions: 1) Exploiting the cyclic symmetry of primitives, the particle in the first position of quark–gluon primitives shall always be a fermion whose fermion flow is pointing into the diagram. 2) We require this fermion to belong to a left-turning fermion-pair which is always possible due to mirror symmetry. With this convention, we need to distinguish only between two separately gauge invariant classes of primitives:

- Primitive amplitudes with mixed quark-gluon loop content $A_n^{[m]}$ (they include also the pure gluonic case which we previously called $A_n^{[1]}$)

- Primitive amplitudes with a closed fermion loop $A_n^{[f]} \equiv A_n^{[1/2]}$.

It is not difficult to convince oneself that different permutations of external particles cover all possible assignments of the handedness discussed above. All propagator flavours of the corresponding parent diagrams $\mathcal{F}_{\text{par}} = [F_1, \ldots, F_n]$ are now uniquely fixed, given

1. A list of external particle flavours $\mathcal{F}_{\text{ext}} = [f_1, \ldots, f_n]$

2. The flavour of the propagator F_1 in front of the first external particle. This propagator is either a gluon for the mixed quark-gluon case $A_n^{[m]}$, or a quark with different flavour from all external quarks for the closed quark loop $A_n^{[f]}$.

Every subsequent propagator is then determined from the previous one, the external particle and a suitable three-point vertex[4]. In case that certain propagators do not exist — a situation which may occur if a quark pair is enclosed in between another quark pair or in case of a fermion loop with external quark lines — we assign a dummy propagator to have always as many propagators as external legs. The assignment of the propagators for the mixed and the closed quark loop case are shown in Fig. 1.2. Roughly speaking, we can say $[f_1, \ldots, f_n] \oplus F_1 \to [F_1, \ldots, F_n]$.

As an illustration we summarise the colour decomposition of one-loop QCD amplitudes with one external quark line plus an arbitrary number of gluons in terms of primitives which has been studied in great detail in Ref. [74]:

$$\mathcal{A}_n^{\text{1-loop}}(\bar{q}, q, 3, \ldots, n) = g_s^n \sum_{j=1}^{n-1} \sum_{\sigma \in S_n / S_{n;j}} \text{Gr}_{n;j}^{(\bar{q}q)}(\sigma_3, \ldots, \sigma_n)\, A_{n;j}(\bar{q}, q, \sigma_3, \ldots, \sigma_n) \quad (1.37)$$

where

$$\text{Gr}_{n;1}^{(\bar{q}q)}(3, \ldots, n) = N_c\, (T^{a_3} \ldots T^{a_n})_{i_2 \bar{i}_1},$$

$$\text{Gr}_{n;2}^{(\bar{q}q)}(3, \ldots, n) = 0,$$

$$\text{Gr}_{n;j}^{(\bar{q}q)}(3, \ldots, j+1, j+2, \ldots, n) = \text{Tr}(T^{a_3} \ldots T^{j+1})\,(T^{a_{j+2}} \ldots T^{a_n})_{i_2 \bar{i}_1},$$

$$\text{Gr}_{n;n-1}^{(\bar{q}q)}(3, \ldots, n) = \text{Tr}(T^{a_3} \ldots T^{a_n})\, \delta_{i_2 \bar{i}_1}.$$

and $S_n / S_{n;j}$ are those permutations which leave the colour structures $\text{Gr}_{n;j}^{(\bar{q}q)}$ invariant. The leading colour partial amplitude $A_{n;1}$ receives only contributions from primitives where the two quarks are adjacent:

$$A_{n;1}(\bar{q}, q, 3, \ldots, n) = A_n^{[m]}(\bar{q}, q, 3, \ldots, n) - \frac{1}{N_c^2}(-1)^n A_n^{[m]}(\bar{q}, n, \ldots, 3, q)$$

$$+ \frac{n_f}{N_c} A_n^{[f]}(\bar{q}, q, 3, \ldots, n). \quad (1.38)$$

Even if we neglected the n_f dependent contributions, $A_{n;1}$ gets contributions from two different primitive amplitudes, in contrast to the pure gluonic case where the primitive amplitudes are precisely the leading colour partial amplitudes. The subleading colour

---

[4]We use the same quark-index-system which is described in appendix B.2 for the tree-level amplitudes.

	$A^{[0]}$	$A^{[m]}$	$A^{[f]}$		$A^{[0]}$	$A^{[m]}$	$A^{[f]}$
$\mathcal{A}_4(g,g,g,g)$	2	3	3	$\mathcal{A}_5(g,g,g,g,g)$	6	12	12
$\mathcal{A}_4(\bar{q}_1,q_1,g,g)$	2	6	1	$\mathcal{A}_5(\bar{q}_1,q_1,g,g,g)$	6	24	6
$\mathcal{A}_4(\bar{q}_1,q_1,\bar{q}_2,q_2)$	1	4	1	$\mathcal{A}_5(\bar{q}_1,q_1,\bar{q}_2,q_2,g)$	3	16	3
$\mathcal{A}_6(g,g,g,g,g,g)$	24	60	60	$\mathcal{A}_7(g,g,g,g,g,g,g)$	120	360	360
$\mathcal{A}_6(\bar{q}_1,q_1,g,g,g,g)$	24	120	33	$\mathcal{A}_7(\bar{q}_1,q_1,g,g,g,g,g)$	120	720	230
$\mathcal{A}_6(\bar{q}_1,q_1,\bar{q}_2,q_2,g,g)$	12	80	13	$\mathcal{A}_7(\bar{q}_1,q_1,\bar{q}_2,q_2,g,g,g)$	60	480	75
$\mathcal{A}_6(\bar{q}_1,q_1,\bar{q}_2,q_2,\bar{q}_3,q_3)$	4	32	4	$\mathcal{A}_7(\bar{q}_1,q_1,\bar{q}_2,q_2,\bar{q}_3,q_3,g)$	20	192	20

Table 1.1.: The number of required primitive amplitudes at tree-level ($A^{[0]}$) and at one-loop order for the mixed ($A^{[m]}$) and fermion loop ($A^{[f]}$) cases.

partial amplitudes $A_{n;j}$ with $j \geq 3$ are constructed from special linear combinations of primitive amplitudes involving also non-adjacent quark–anti-quark configurations [74]. Both the numerical accuracy and the runtime performance of primitives are sensitive to the quark–anti-quark separations as will be discussed in sections 3.1 and 3.2. We mention that this process has for the n_f-independent part a colour decomposition which mixes both adjoint F and fundamental T matrices [106].

For one-loop QCD amplitudes with more than one external quark-line, no published analytic formula to express all partial amplitudes in terms of primitives is so far available. For these cases, we employ the technique outlined in section 1.2 to map the primitive amplitudes to a set of Feynman diagrams [75, 76, 80]. The number of required primitive amplitudes for the basis channels of 2-jet, 3-jet, 4-jet and 5-jet production both at tree-level and at one-loop order is summarised in table 1.1. The like-flavour amplitudes are obtained upon anti-symmetrisation of the unequal quark indices and therefore contain a larger bases of primitives. The physical channels are obtained exploiting crossing symmetry.

The bottom line from this section is the following: An efficient colour management allows to express the full colour one-loop QCD amplitudes exclusively in terms of primitive amplitudes. The one-loop primitive amplitudes are colour stripped objects with a much simpler kinematic structure than the full amplitude. They can be represented by a parent diagram which specifies uniquely the particle content within the loop by means of a flavour list $\mathcal{F}_{\mathrm{par}} = [F_1, \ldots, F_n]$. The numerical computation of primitive amplitudes by means of integrand reduction techniques will be subject of chapter 2. The parent diagram will thereby play the rôle of the backbone of the computation.

1.4. Colour summed squared helicity amplitudes

The full colour Born amplitudes are either squared or interfered with the full colour one-loop amplitudes, followed by a summation over the external colours. The obtained colour summed squared helicity amplitudes are implemented in NJET for all channels of 2-jet, 3-jet, 4-jet and 5-jet production and can be computed for arbitrary helicities of the external partons.

We use the simple five point tree-level example $\mathcal{A}(q, \bar{q}, Q, \bar{Q}, g)$ in Eq. 1.33 from section 1.2 to illustrate the colour summation of helicity amplitudes. We first collect the independent tree-level primitives in a vector $\vec{\mathcal{P}}^{(0)T} = (\mathcal{P}^{(0)}_{[12345]}, \mathcal{P}^{(0)}_{[12354]}, \mathcal{P}^{(0)}_{[12534]})$ and write the amplitude as

$$
\mathcal{A}(q, \bar{q}, Q, \bar{Q}, g) \equiv \vec{X}^{(0)} \cdot \vec{\mathcal{P}}^{(0)} = \left((T^{a_5})_{i_1 \bar{i}_4} \delta_{i_3 \bar{i}_2} - \frac{1}{N_c} (T^{a_5})_{i_1 \bar{i}_2} \delta_{i_3 \bar{i}_4} \right) \mathcal{P}^{(0)}_{[12345]}
$$
$$
+ \left(\frac{1}{N_c} \delta_{i_1 \bar{i}_2} (T^{a_5})_{i_3 \bar{i}_4} - \frac{1}{N_c} (T^{a_5})_{i_1 \bar{i}_2} \delta_{i_3 \bar{i}_4} \right) \mathcal{P}^{(0)}_{[12354]}
$$
$$
+ \left(\delta_{i_1 \bar{i}_4} (T^{a_5})_{i_3 \bar{i}_2} - \frac{1}{N_c} (T^{a_5})_{i_1 \bar{i}_2} \delta_{i_3 \bar{i}_4} \right) \mathcal{P}^{(0)}_{[12534]}
$$

where the components of the colour vector $\vec{X}^{(0)}$ can directly be read off as the terms multiplying the components $\mathcal{P}^{(0)}_i$. The colour summed squared helicity amplitude may thus be written as

$$
\mathrm{d}\sigma^{\mathrm{Born}} \sim \sum_{\mathrm{colour}} \left| \mathcal{A}(q, \bar{q}, Q, \bar{Q}, g) \right|^2 = \vec{\mathcal{P}}^{(0)\dagger} \, \hat{\mathcal{C}}^{(0)} \, \vec{\mathcal{P}}^{(0)} \tag{1.39}
$$

where the Born level colour matrix $\hat{\mathcal{C}}^{(0)}$ is defined as

$$
\hat{\mathcal{C}}^{(0)}_{ij} = \sum_{\mathrm{colour}} X_i^{(0)} X_j^{(0)}. \tag{1.40}
$$

For the example $\mathcal{A}(q, \bar{q}, Q, \bar{Q}, g)$, the explicit form of $\hat{\mathcal{C}}^{(0)}$ reads

$$
\hat{\mathcal{C}}^{(0)} = \frac{N_c^2 - 1}{N_c} \begin{pmatrix} (N_c^2 - 1) & 1 & -1 \\ 1 & 2 & 1 \\ -1 & 1 & (N_c^2 - 1) \end{pmatrix}.
$$

This procedure is very general and works in complete analogy also at one-loop order. The one-loop interference term with the Born amplitude reads

$$
\mathrm{d}\sigma^{\mathrm{virtual}} \sim 2 \, \mathrm{Re} \left[\vec{\mathcal{P}}^{(0)\dagger} \, \hat{\mathcal{C}}^{(1)} \, \vec{\mathcal{P}}^{(1)} \right]. \tag{1.41}
$$

$\mathcal{P}^{(0)}$ denotes tree-level primitives while $\mathcal{P}^{(1)}$ are one-loop primitive amplitudes. Since there are in general more colour structures at one-loop order than at tree-level, and, thus, also more primitive amplitudes at one-loop order than at Born level, the colour matrix $\hat{\mathcal{C}}^{(1)}$ is in general not square any more.

2. Integrand reduction techniques at one-loop order

This chapter is dedicated to the computation of arbitrary one-loop primitive amplitudes in massless QCD with unitarity methods and integrand reduction techniques. In
section 2.1, we first discuss well known properties of one-loop amplitudes including
their decomposition in terms of scalar master integrals, their universal pole structure
and the splitting of the amplitude in a cut-constructible and a rational part. In sections 2.2 and 2.3, we review the general tensor structure of the one-loop integrand
using the van Neerven-Vermaseren basis and explain, how the integrand can be reconstructed with help of on-shell tree-level amplitudes. This serves as a starting point to
describe in section 2.4 a method to compute the rational part of massless QCD primitives in strictly four space-time dimensions. Sections 2.5 and 2.6 show the explicit
integrand parametrisations in terms of orthogonal functions to compute both the cut-
constructible part and the rational part. Finally, in section 2.7, we discuss some pecu-
liarities of the tree-level amplitudes that feed into the unitarity cuts.

2.1. The general structure of one-loop amplitudes

2.1.1. Kinematical conventions

To fix the kinematical conventions of a one-loop n-point primitive amplitude, we consider a parent diagram (c.f. section 1.3) where all flavour information is suppressed as
depicted in Fig. 2.1. We take all n external momenta as outgoing and label the particles clockwise from 1 to n. The momentum flow inside the loop is also assumed to be
clockwise. The propagators are labeled such that the mth propagator carries the same
index as the subsequent outgoing mth particle with momentum p_m. We introduce an
additional label R to denote the topology of the occurring scalar master integrals of
the amplitude. For the specific values $R \in \{1, 2, 3, 4, 5\}$, we will refer to $R = 5$ as the
pentagon topology, $R = 4$ as the box topology, $R = 3$ as the triangle topology, $R = 2$
as the bubble topology and $R = 1$ as the tadpole topology. Yet, to understand the general structures it is more instructive to leave R as a free parameter. Depending on the
topology R we consider always a set $C_R = \{i_1, i_2, \ldots, i_R\}$ of propagator labels where
$1 \leq i_1 < i_2 < \ldots < i_R \leq n$. In the following, we will call both R and C_R topology,
it should become clear from the context what is actually meant. The numbers in C_R

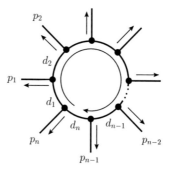

Figure 2.1.: Kinematical conventions for primitive amplitudes: The arrows denote the momentum flow, p_i the external particles' outgoing momenta and d_i the inverse propagators defined in Eq. (2.3).

define a set of R outflow momenta $\mathcal{K}_R = \{k_1, k_2, \ldots, k_R\}$ with

$$k_r = \sum_{m=i_r}^{i_{r+1}-1} p_m \,. \tag{2.1}$$

The indices r are cyclic in R such that $i_{R+1} = i_1$, and the sum over m is cyclic in n such that $\sum_{i_R}^{i_1-1} = \sum_{i_R}^{n} + \sum_{1}^{i_1-1}$. It follows immediately from four-momentum conservation that

$$\sum_{m=1}^{n} p_m = \sum_{r=1}^{R} k_r = 0. \tag{2.2}$$

The inverse propagators are defined as

$$d_{i_r} = d_{i_r}(\ell) = (\ell - q_r)^2 - m_{i_r}^2 = \ell_r^2 - m_{i_r}^2 \tag{2.3}$$

ℓ is the loop momentum and m_{i_r} is the mass of the propagating particle. q_r is the internal momentum parametrisation of a distinct topology C_R and is determined up to a constant momentum q_0 via the outflow momenta k_r as

$$q_{r+1} = q_r + k_r. \tag{2.4}$$

We choose q_0 always such that $q_2 = 0$. This fixes uniquely all q_r according to

$$
\begin{aligned}
q_1 &= -k_1 & \ell_1 &= \ell + k_1 \\
q_2 &= 0 & \ell_2 &= \ell \\
q_3 &= k_2 & \ell_3 &= \ell - k_2 \\
&\;\;\vdots & &\;\;\vdots \\
q_R &= \sum_{r=2}^{R-1} k_r & \ell_R &= \ell - \sum_{r=2}^{R-1} k_r
\end{aligned}
\tag{2.5}
$$

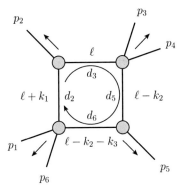

Figure 2.2.: Example of a box topology ($R = 4$) where $C_4 = \{2,3,5,6\}$ and $n = 6$. The arrows indicate the momentum flow. The loop momentum parametrisation is chosen according to Eq. (2.5) as $k_1 = p_2$, $k_2 = p_3 + p_4$, $k_3 = p_5$ and $k_4 = p_6 + p_1$. The inverse propagators d_{i_r} are labelled as they appear in the parent diagram.

ℓ_r is the momentum in the inverse propagator d_{i_r}. Hence, when we talk about the loop momentum ℓ, with our parametrisation this is always the momentum as it appears in the second inverse propagator d_{i_2} from the list $[d_{i_1}, d_{i_2}, \ldots, d_{i_R}]$. An example is given in Fig. 2.2 for a box topology ($R = 4$). With these conventions, a scalar one-loop integral reads

$$\mathcal{I}_{R;i_1\ldots i_R} = \int \frac{\mathrm{d}^D\ell}{i\pi^{D/2}} \frac{1}{d_{i_1}\ldots d_{i_R}} \tag{2.6}$$

where dimensional regularisation is assumed with the convention $D = 4 - 2\epsilon$.

2.1.2. Scalar integral basis

In a traditional approach based on Feynman diagrams, the computation of a one-loop amplitude works as follows: One writes down all one-loop Feynman diagrams contributing to a distinct process. Every diagram is then translated into algebraic expressions which depend on external momenta and sources (spinors for fermions and polarisation vectors for vector bosons), masses and on the loop momentum. Subsequently, the tensor loop integrals are reduced to a set of master integrals, for example by means of a Passarino-Veltman (PV) reduction [16]. As a result of the PV reduction, any Feynman diagram and, hence, any amplitude can be written as a linear combination of scalar one-point, two-point, three-point and four-point integrals

$$\mathcal{A} = \sum_{[i_1|i_4]} d^{[D]}_{i_1i_2i_3i_4} \mathcal{I}_{4;i_1i_2i_3i_4} + \sum_{[i_1|i_3]} c^{[D]}_{i_1i_2i_3} \mathcal{I}_{3;i_1i_2i_3} + \sum_{[i_1|i_2]} b^{[D]}_{i_1i_2} \mathcal{I}_{2;i_1i_2} + \sum_{[i_1|i_1]} a^{[D]}_{i_1} \mathcal{I}_{1;i_1}. \tag{2.7}$$

$[i_1|i_R]$ denotes a sum over all combinations of indices i_r for which $1 \leq i_1 < i_2 < \ldots < i_R \leq n$ holds. The coefficients in front of every integral are functions of Lorentz invariants depending only on the external kinematics, polarisations and both external and internal masses. It is important to stress that the coefficients are in $D = 4 - 2\epsilon$ dimensions, hence, they may also explicitly depend on the dimensional regularisation parameter ϵ. We indicated this fact through the superscript $[D]$ at the integral coefficients.

The scalar integrals are functions of the external momenta and of the propagator masses. They are all known analytically. In particular, there exist various public libraries which collect numerical implementations of the integrals like for example FF/QCDLOOP [117, 118], LOOPTOOLS [119], ONELOOP [120], GOLEM95(C) [121, 122]. Since we consider the case of massless QCD, the tadpole integrals $\mathcal{I}_{1;i}$ are scale free and, therefore, are set equal to zero in dimensional regularisation. Expanding in the dimensional regularisation parameter ϵ and taking the limit $\epsilon \rightarrow 0$, the integral coefficients become four-dimensional, however, this generates additional rational terms which are related to the ultra violet (UV) behaviour of the amplitude.

$$\mathcal{A} = \sum_{[i_1|i_4]} d_{i_1 i_2 i_3 i_4}^{[4]} \, \mathcal{I}_{4;i_1 i_2 i_3 i_4} + \sum_{[i_1|i_3]} c_{i_1 i_2 i_3}^{[4]} \, \mathcal{I}_{3;i_1 i_2 i_3} + \sum_{[i_1|i_2]} b_{i_1 i_2}^{[4]} \, \mathcal{I}_{2;i_1 i_2} + \mathcal{R}. \qquad (2.8)$$

In particular, the rational terms arise, if terms of order ϵ from the integral coefficients are multiplied with the $\frac{1}{\epsilon}$-UV poles inside the master integrals. It can be shown that infrared (IR) poles do not lead to rational terms [123]. Since the scalar box and triangle integrals have only IR poles and are all UV finite, only the scalar bubbles will contribute to the rational part.

The PV reduction can become quite lengthy, especially for the case of higher multiplicities. Knowing the general one-loop decomposition in Eq. (2.8), the natural question arises whether it is possible to compute the integral coefficients and the rational terms directly without performing the reduction for every single diagram. We show explicitly in sections 2.3 and 2.4 that the new methods based on unitarity and integrand reduction developed in the past few years provide successful tools for a direct computation of both every individual integral coefficient and the rational part.

2.1.3. Comments on the rational part

For certain classes of amplitudes, the rational part vanishes. At the diagrammatic level, the rational part in a Feynman m-point integral of tensor rank r vanishes if the following condition is satisfied [55]:

$$r < \max\{(m - 1), 2\} \qquad (2.9)$$

in other words, the tensor rank must always be two units lower than the multiplicity of the integral, except for the bubbles for which a difference of one unit is sufficient. Naively, one expects that a UV finite integral does not have a rational part (as we argued above with the scalar boxes and triangles). At the level of tensor integrals, this is not necessarily true any more. For example, according to Eq. 2.9, the UV finite rank-three four-point integral does have a rational part, while the UV-finite rank-two four-point integral does not. Those amplitudes in which all Feynman integrals obey the power counting criterion from Eq. (2.9) are called *cut-constructible*. In Ref. [55] it is shown that cut-constructible amplitudes can unambiguously be constructed from the knowledge of branch cuts in the complex plane. The rough idea behind this approach is the following: In the Feynman integrals fulfilling Eq. (2.9), the logarithms of momentum invariants appear in such a combination that every integral can be uniquely associated with at least one distinct type of logarithm or products thereof. If one analytically continues the momentum invariants in the complex plane, the logarithms exhibit branch cuts. These branch cuts can be related via the Cutkosky rules [56] to a product of tree-level amplitudes which was the original motivation of the unitarity method. The outlined ideas go already back to the sixties [56, 124–126]. While it can be shown that for example $\mathcal{N} = 4$ and $\mathcal{N} = 1$ Super-Yang-Mills theory are entirely cut-contstructible [55], QCD is definitively not. It does not obey the power counting criterion in Eq. (2.9) and, hence, we have to explicitly compute both the cut-constructible and the rational part. Nevertheless, we will make extensive use of the power counting criterion in section 2.4 when we introduce the methods to compute the rational part.

2.1.4. Scheme dependence and universal pole structure

Here, all computations at the level of primitive amplitudes are performed in the four-dimensional helicity (FDH) scheme [127]. In this scheme, all external momenta and external polarisations and the polarisations of the particles circulating in the loop are four-dimensional. The loop momentum, however is taken in $D = 4 - 2\epsilon$ dimensions. Since it can be shown that the FDH scheme is equivalent to dimensional reduction (DR) at one-loop order [128], the conversion to conventional dimensional regularisation (CDR) is straightforward and gives rise to finite terms proportional to the Born amplitude. Its explicit form including the analytic expressions for the universal UV and IR poles of the amplitude are given in Ref. [129]. The correct reproduction of the poles represents a strong check of the amplitude's validity as will be discussed in section 3.1. The poles of the full colour summed amplitude read

$$
\text{Re}\left[\mathcal{A}^{(0)\dagger} \cdot \mathcal{A}_U^{(1),FDH}\right] = \frac{1}{\epsilon^2}\left(C_A\, n_g + C_F\, n_q\right) g_s^2 |\mathcal{A}^{(0)}|^2
$$
$$
+ \sum_{i=1}^{n-1}\sum_{j=i+1}^{n} \frac{1}{\epsilon}\log\left(\frac{\mu_R^2}{-s_{ij}}\right) g_s^2 |\mathcal{A}_{ij}^{(0)}|^2 - \frac{1}{\epsilon}\left(\beta_0 - n_q\left(\frac{\beta_0}{2} - \frac{3}{2}C_F\right)\right) g_s^2 |\mathcal{A}^{(0)}|^2
$$
$$
+ \text{ finite terms.} \tag{2.10}
$$

$\mathcal{A}_U^{(1),FDH}$ denotes the unrenormalised one-loop amplitude in the FDH scheme, n_g is the number of external gluons, n_q the number of external quarks and the Casimir operators read $C_A = N_c$, $C_F = (N_c^2 - 1)/2N_c$. g_s is the strong coupling constant and $\beta_0 =$

$(11C_A - 2n_f)/3$ the first coefficient of the QCD beta function. $|\mathcal{A}^{(0)}|^2$ is the (full colour) Born amplitude squared. The object $|\mathcal{A}_{ij}^{(0)}|^2$ is the colour correlated Born amplitude squared defined as

$$|\mathcal{A}_{ij}^{(0)}|^2 = \langle \mathcal{A}^{(0)\dagger} | T_i T_j | \mathcal{A}^{(0)} \rangle \qquad (2.11)$$

with $T_g = f^{abc}$ and $T_q = T_{i_1 \bar{i}_2}^a$. We have used the same notation as in [129] to denote by $|\mathcal{A}^{(0)}\rangle$ the leading order amplitude. The bra-ket notation suggests that this can be understood as a vector in colour space. The $\frac{1}{\epsilon^2}$-poles and the poles proportional to the logarithms are of pure infrared origin. The remaining $\frac{1}{\epsilon}$-poles are both UV and IR poles. To disentangle the contributions, it is helpful to consider the difference of charge renormalised and unrenormalised quantities

$$\frac{\mathrm{Re}\left[\mathcal{A}^{(0)\dagger} \cdot \left(\mathcal{A}_R^{(1),FDH} - \mathcal{A}_U^{(1),FDH}\right)\right]}{g_s^2 |\mathcal{A}^{(0)}|^2} = -\frac{(n_g + n_q - 2)}{2}\left(\frac{\beta_0}{\epsilon} - \frac{1}{3}C_A\right). \qquad (2.12)$$

Note the scheme dependent finite contribution $\sim C_A$ which is specific for the FDH/DR scheme and which vanishes in the CDR scheme, c.f. Ref. [129] and the cited references therein. For convenience, we cite also the scheme conversion between FDH/DR and CDR.

$$\frac{\mathrm{Re}\left[\mathcal{A}^{(0)\dagger} \cdot \left(\mathcal{A}_U^{(1),CDR} - \mathcal{A}_U^{(1),FDH}\right)\right]}{g_s^2 |\mathcal{A}^{(0)}|^2} = \frac{(n_g + n_q - 2)}{2}\frac{1}{3}C_A - n_g\frac{1}{6}C_A - n_q\frac{1}{2}C_F. \qquad (2.13)$$

To conclude this section we give a heuristic derivation how to check also the poles of individual primitive amplitudes. We derive our formulae from Eq. (9) of Ref. [129]. First we disentangle the n_f dependent contributions from the rest. This gives the pole function of a primitive with closed quark loop

$$A_{\mathrm{pole}}^{[f]} = -\frac{1}{3\epsilon}(2 - \hat{n}_q)A^{[0]} \qquad (2.14)$$

where we explicitly set $n_f = 1$ and $\hat{n}_q$ denotes the number of external quark lines which may couple directly via a gluon propagator to the fermion loop. Those external fermion lines which are enclosed between other external fermion lines cannot couple directly to the fermion loop and are explicitly excluded from $\hat{n}_q$. (C.f. the discussion in section 1.3 on primitive topologies with enclosed external fermion lines.) $A^{[0]}$ is the corresponding tree-level primitive amplitude. For the soft gluon contribution we start from the purely gluonic case and discard all contributions with non-adjacent momenta as stated in [74]

$$V_g = -\frac{1}{\epsilon^2}\sum_{i=1}^{n}\left(\frac{\mu^2}{-s_{i,i+1}}\right)^{\epsilon}. \qquad (2.15)$$

For the mixed quark-gluon case, we have to remove all contributions from adjacent momenta which enclose a quark propagator in the parent diagram. For example, for the primitive amplitude $A^{[m]}(\bar{q}_1, q_1, \bar{q}_2, q_2, 5, \ldots, n)$, this would be $V_g + 1/\epsilon^2[(\mu^2/s_{12})^\epsilon + (\mu^2/s_{34})^\epsilon]$. The remaining constant terms we get from Ref. [129] expanding Eq. (9) in leading colour and omitting the soft gluon and the n_f contributions. Putting all ingredients together, the final formula reads

$$A^{[m]}_{\text{pole}} = \left[-\frac{1}{\epsilon^2} \sum_{i \in n_g} \left(\frac{\mu^2}{-s_{i,i+1}} \right)^\epsilon + \frac{1}{\epsilon} \left(-\frac{11}{3} + \frac{13}{12} \hat{n}_q \right) \right] A^{[0]}. \tag{2.16}$$

The suggestive notation $i \in n_g$ means subtracting the quarks in the parent diagram as explained above. $\hat{n}_q$ excludes again enclosed external fermion lines. For $\hat{n}_q = 0$, this formula reproduces the one given in Ref. [77]. We could confirm Eqs. (2.14) and (2.16) numerically in all examples that we considered (up to 20 external legs and 10 fermion lines).

2.2. The van Neerven-Vermaseren basis

We introduce now a specific basis in D-dimensional Minkowski space which goes back to van Neerven and Vermaseren (NV) in the context of tensor reduction [17]. The construction is explained in detail in the paper from Ellis, Giele and Kunszt [72]. We review their arguments and specify some additional details. Expressing the loop-momentum in terms of the NV-basis intents to meet the following two important objectives: First, to derive the general form of the integrand of a one-loop amplitude as will be shown in section 2.3. Second, to impose on-shell constraints on the loop momentum being necessary in sections 2.5 and 2.6.

We assume that in D integer space-time dimensions, R outflow momenta $k_1, \ldots, k_R$ for which momentum conservation $\sum_{r=1}^R k_r = 0$ holds are given as described in section 2.1.1. The dimension of the *physical space* that is spaned by the R momenta is therefore $D_P = \min(D, R-1)$. For $R \leq D$, the physical space is a lower dimensional subspace P of the Minkowski space M. The orthogonal complement T with $P + T = M$ is called *transverse space*[1]. In particular holds

$$D = D_P + D_T, \tag{2.17}$$

$$D_P = \min(D, R-1), \tag{2.18}$$

$$D_T = \max(0, D-R+1). \tag{2.19}$$

The NV-basis is a basis in Minkowski space with D_P vectors v_i^μ in the physical space

---

[1] In Ref. [72], the transverse space is called trivial space.

and D_T orthonormal vectors n_i^μ in the transverse space that fulfil the properties

$$v_i \cdot k_j = \delta_{ij}, \tag{2.20}$$

$$n_i \cdot n_j = \delta_{ij}, \tag{2.21}$$

$$n_i \cdot k_j = n_i \cdot v_j = 0. \tag{2.22}$$

Note that in general $v_i \cdot v_j \neq 0$. We sketch now briefly how these vectors can be constructed explicitly.

The generalised Kronecker-delta in a D-dimensional Minkowski space with $R \leq D$ is defined as:

$$\delta_{\nu_1\nu_2\cdots\nu_R}^{\mu_1\mu_2\cdots\mu_R} = \begin{vmatrix} \delta_{\nu_1}^{\mu_1} & \delta_{\nu_2}^{\mu_1} & \cdots & \delta_{\nu_R}^{\mu_1} \\ \delta_{\nu_1}^{\mu_2} & \delta_{\nu_2}^{\mu_2} & \cdots & \delta_{\nu_R}^{\mu_2} \\ \vdots & \vdots & & \vdots \\ \delta_{\nu_1}^{\mu_R} & \delta_{\nu_2}^{\mu_R} & \cdots & \delta_{\nu_R}^{\mu_R} \end{vmatrix}, \qquad \mu_i, \nu_j = 0, \dots, D-1. \tag{2.23}$$

This function is totally anti-symmetric in its upper and lower indices. For $R > D$, the generalised Kronecker delta equals zero. In the special case in which $R = D$, the generalised Kronecker delta factorises into a product of two Levi-Civita epsilon tensors:

$$\delta_{\nu_1\nu_2\cdots\nu_D}^{\mu_1\mu_2\cdots\mu_D} \sim \varepsilon_{\nu_1\nu_2\cdots\nu_D}\varepsilon^{\mu_1\mu_2\cdots\mu_D}. \tag{2.24}$$

The $(R-1)$-particle Gram determinant $\Delta(k_1, k_2, \dots, k_{R-1})$ is defined as the contraction of $(R-1)$ outflow momenta k_i with the generalised Kronecker delta

$$\Delta(k_1, k_2, \dots, k_{R-1}) \equiv \delta_{k_1 k_2 \dots k_{R-1}}^{k_1 k_2 \dots k_{R-1}} \tag{2.25}$$

where we have used the shorthand notation

$$\delta_{P\nu_2\cdots\nu_R}^{\mu_1 Q\cdots\mu_R} \equiv \delta_{\nu_1\nu_2\cdots\nu_R}^{\mu_1\mu_2\cdots\mu_R} P^{\nu_1} Q_{\mu_2}. \tag{2.26}$$

The physical vectors v_i are then given by

$$v_i^\mu(k_1, \dots, k_{R-1}) = \frac{\delta_{k_1\dots k_{i-1}k_i k_{i+1}\dots k_{R-1}}^{k_1\dots k_{i-1}\mu k_{i+1}\dots k_{R-1}}}{\Delta(k_1, \dots, k_{R-1})}. \tag{2.27}$$

With the properties of the generalised Kronecker delta, one sees immediately that $v_i \cdot k_j = \delta_{ij}$ is fulfilled as required from the defining Eq. (2.20). In order to compute the transverse vectors, a projection operator onto the transverse space is defined as

$$w_\mu{}^\nu = \frac{\delta_{k_1 k_2 \dots k_{R-1}\mu}^{k_1 k_2 \dots k_{R-1}\nu}}{\Delta(k_1, \dots, k_{R-1})}. \tag{2.28}$$

This operator satisfies the following identities which are proven in appendix C:

$$w_\mu{}^\mu = D_T = D + 1 - R, \qquad (2.29)$$

$$k_i^\mu w_\mu{}^\nu = v_i^\mu w_\mu{}^\nu = 0, \qquad (2.30)$$

$$w^{\mu\alpha} w_\alpha{}^\nu = w^{\mu\nu}. \qquad (2.31)$$

The last equation shows that the projection operator $w^{\mu\nu}$ has the properties of a metric tensor of the transverse subspace. We can thus decompose $w^{\mu\nu}$ in terms of the orthonormal basis vectors n_i^μ that span the transverse subspace

$$w^{\mu\nu} = \sum_{i=1}^{D_T} n_i^\mu n_i^\nu. \qquad (2.32)$$

How to explicitly construct the basis vectors n_i^μ such that Eqs. (2.21) and (2.22) are fulfilled is explained below. With these ingredients, we can expand the metric tensor $g^{\mu\nu}$ in terms of the basis vectors. We start with the identity $D = D_P + D_T$ given in Eq. (2.17) and apply Eqs. (2.20), (2.29) and (2.32):

$$D = D_P + D_T$$

$$g_{\mu\nu} g^{\mu\nu} = \sum_{i,j=1}^{D_P} \delta_{ij} + w_\nu{}^\nu = \sum_{i=1}^{D_P} k_i \cdot v_i + g_{\mu\nu} w^{\mu\nu} = g_{\mu\nu} \left(\sum_{i=1}^{D_P} k_i^\mu v_i^\nu + \sum_{i=1}^{D_T} n_i^\mu n_i^\nu \right)$$

Comparing the left-hand side and the right-hand side, we arrive at the decomposition of the metric tensor

$$g^{\mu\nu} = \sum_{i=1}^{D_P} k_i^\mu v_i^\nu + \sum_{i=1}^{D_T} n_i^\mu n_i^\nu. \qquad (2.33)$$

Eq. (2.33) is indeed symmetric in μ and ν because $\sum_{i=1}^{D_P} k_i^\mu v_i^\nu = \sum_{i=1}^{D_P} v_i^\mu k_i^\nu$ which follows from the above derivation.

For our application we will need the case $D = 4$ and $D = 5$. The latter will be necessary to incorporate the dependence of the dimensional regularisation parameter ϵ which is present in the loop momentum. We will first construct all basis vectors for the four-dimensional case, subsequently we will see that in our situation, only a tiny modification is needed to cover also the five-dimensional case.

2.2.1. Construction of the basis vectors in four space-time dimensions

The relation between the topology and the dimension of physical and transverse space-time is summarised in table 2.1 for the case $D = 4$. Recall that the dimension of the physical space coincides with the number of independent outflow momenta.

topology	R	D_P	D_T
pentagons	5	4	0
boxes	4	3	1
triangles	3	2	2
bubbles	2	1	3

Table 2.1.: Relation between topology $R \in \{2, 3, 4, 5\}$ and the dimension of physical and transverse space in $D = 4$ dimensions. The dimension of the physical space coincides with the number of independent outflow momenta.

Pentagon basis: For $R = 5$ and $D = 4$, we have four independent outflow momenta k_1, k_2, k_3 and k_4, hence, for the NV basis we need to construct four vectors v_i in the physical space. There are no vectors in the transverse space. With Eq. (2.27), the basis in terms of the v_i reads:

$$v_1^\mu(k_1, k_2, k_3, k_4) = \frac{\delta_{k_1 k_2 k_3 k_4}^{\mu k_2 k_3 k_4}}{\Delta(k_1, k_2, k_3, k_4)}, \qquad v_2^\mu(k_1, k_2, k_3, k_4) = \frac{\delta_{k_1 k_2 k_3 k_4}^{k_1 \mu k_3 k_4}}{\Delta(k_1, k_2, k_3, k_4)},$$

$$v_3^\mu(k_1, k_2, k_3, k_4) = \frac{\delta_{k_1 k_2 k_3 k_4}^{k_1 k_2 \mu k_4}}{\Delta(k_1, k_2, k_3, k_4)}, \qquad v_4^\mu(k_1, k_2, k_3, k_4) = \frac{\delta_{k_1 k_2 k_3 k_4}^{k_1 k_2 k_3 \mu}}{\Delta(k_1, k_2, k_3, k_4)}. \qquad (2.34)$$

Box basis: For $R = 4$ and $D = 4$, we have three independent outflow momenta k_1, k_2 and k_3, hence, for the NV basis we need to construct three vectors v_i in the physical space and one transverse space vector n_1. With Eq. (2.27), the basis in terms of the v_i reads:

$$v_1^\mu(k_1, k_2, k_3) = \frac{\delta_{k_1 k_2 k_3}^{\mu k_2 k_3}}{\Delta(k_1, k_2, k_3)}, \qquad v_2^\mu(k_1, k_2, k_3) = \frac{\delta_{k_1 k_2 k_3}^{k_1 \mu k_3}}{\Delta(k_1, k_2, k_3)},$$

$$v_3^\mu(k_1, k_2, k_3) = \frac{\delta_{k_1 k_2 k_3}^{k_1 k_2 \mu}}{\Delta(k_1, k_2, k_3)}. \qquad (2.35)$$

In this special case in which $D = R = 4$, the transverse vector n_1^μ is simply given by the factorisation property of the generalised Kronecker delta into two Levi-Civita tensors, as explained in Eq. (2.24). We use the sign conventions for the four-dimensional Levi-Civita tensor in Minkowski space given in Ref. [130]. This also determines the sign in Eq. (2.24)

$$\varepsilon^{0123} = -\varepsilon_{0123} = 1, \qquad (2.36)$$

$$\delta_{\nu_1 \nu_2 \nu_3 \nu_4}^{\mu_1 \mu_2 \mu_3 \mu_4} = -\varepsilon_{\nu_1 \nu_2 \nu_3 \nu_4} \varepsilon^{\mu_1 \mu_2 \mu_3 \mu_4}. \qquad (2.37)$$

Note that this convention differs from the one given in Ref. [72]. Thus, we write

$$w_\mu^{\ \nu} = n_{1\mu} n_1^\nu = \frac{\delta_{k_1 k_2 k_3 \mu}^{k_1 k_2 k_3 \nu}}{\Delta(k_1, k_2, k_3)} = -\frac{\varepsilon_{k_1 k_2 k_3 \mu} \varepsilon^{k_1 k_2 k_3 \nu}}{\Delta(k_1, k_2, k_3)}. \tag{2.38}$$

Since the transverse vectors are normalised according to $n_1^2 = 1$, n_1^μ is determined up to an overall sign:

$$n_1^\mu = \frac{\pm i\, \varepsilon^{k_1 k_2 k_3 \mu}}{\sqrt{\Delta(k_1, k_2, k_3)}}. \tag{2.39}$$

From the two possible solutions we choose the one with positive sign.

Triangle basis: For $R = 3$ and $D = 4$, we have two independent outflow momenta k_1 and k_2, hence, for the NV basis we need to construct two vectors v_i in the physical space and two vectors n_i in the transverse space. With Eq. (2.27), the basis in terms of the v_i reads:

$$v_1^\mu(k_1, k_2) = \frac{\delta_{k_1 k_2}^{\mu k_2}}{\Delta(k_1, k_2)}, \qquad v_2^\mu(k_1, k_2) = \frac{\delta_{k_1 k_2}^{k_1 \mu}}{\Delta(k_1, k_2)}. \tag{2.40}$$

Since $R \neq D$, the generalised Kronecker delta does not factorise. In order to get two linearly independent vectors in the transverse space, we follow the suggestions given in Ref. [87]. The basic idea is to choose an auxiliary vector b_1 which is not collinear to the outflow momenta k_1 and k_2 and to consider this vector as a fictitious "pseudo" outflow momentum. The projection of this vector onto the transverse space is proportional to a first transverse vector $n_1^\nu \sim w_\mu^{\ \nu} b_1^\mu$, and the contraction with the Levi-Civita tensor to a second one $n_2^\mu \sim \varepsilon^{\mu b_1 k_1 k_2}$. This procedure can be generalised to D-dimensional space-time with a $(D - R + 1)$-dimensional transverse subspace. Given $(R - 1)$ outflow momenta $\{k_1, \ldots, k_{R-1}\}$, one chooses $(D - R + 1)$ auxiliary vectors $\{b_1, \ldots, b_{D-R+1}\}$ which are not collinear to the outflow momenta. A transverse vector n_i^μ with $n_i \cdot n_j = \delta_{ij}$ and $v_i \cdot n_j = k_i \cdot n_j = 0$ is then given by

$$n_i^\mu = \frac{\delta_{b_i b_{i-1} b_1 k_1 \ldots k_{R-1}}^{\mu b_{i-1} b_1 k_1 \ldots k_{R-1}}}{\sqrt{\Delta(b_{i-1}, \ldots, b_1 k_1, \ldots, k_{R-1}) \Delta(b_i, \ldots, b_1 k_1, \ldots, k_{R-1})}}. \tag{2.41}$$

Although one chooses $(D - R + 1)$ auxiliary vectors, only $(D - R)$ vectors are needed because the dependence on b_{D-R+1} in the last step drops out due to the factorisation of the generalised Kronecker delta into two Levi-Civita tensors. For our case of the triangle basis, the two vectors read

$$n_1^\mu = \frac{\delta_{b_1 k_1 k_2}^{\mu k_1 k_2}}{\sqrt{\Delta(k_1, k_2)\Delta(b_1, k_1, k_2)}}, \tag{2.42}$$

$$n_2^\mu = \frac{\delta_{b_2 b_1 k_1 k_2}^{\mu b_1 k_1 k_2}}{\sqrt{\Delta(b_1, k_1, k_2)\Delta(b_2, b_1, k_1, k_2)}} = \frac{i\, \varepsilon^{\mu b_1 k_1 k_2}}{\sqrt{\Delta(b_1, k_1, k_2)}}. \tag{2.43}$$

The construction of n_2^μ works formally in an analogous way as the single transverse box vector in Eq. (2.39) which is obvious if one considers the momenta b_1, k_1 and k_2 as outflow momenta.

Bubble basis: For $R = 2$ and $D = 4$, we have only one independent outflow momentum k_1 hence, for the NV basis we need to construct one vector v_i in the physical space and three vectors n_i in the transverse space. With Eq. (2.27), v_1 simply reads:

$$v_1^\mu(k_1) = \frac{\delta_{k_1}^\mu}{\Delta(k_1)} = \frac{k_1^\mu}{k_1^2}. \tag{2.44}$$

For the transverse basis vectors $\{n_1, n_2, n_3\}$ we apply again the master formula Eq. (2.41) which we have described in the previous paragraph. With the three auxiliary vectors b_1, b_2 and b_3 the transverse vectors are given by

$$n_1^\mu = \frac{\delta_{b_1 k_1}^{\mu k_1}}{\sqrt{\Delta(k_1)\Delta(b_1, k_1)}}, \tag{2.45}$$

$$n_2^\mu = \frac{\delta_{b_2 b_1 k_1}^{\mu b_1 k_1}}{\sqrt{\Delta(b_1, k_1)\Delta(b_2, b_1, k_1)}}, \tag{2.46}$$

$$n_3^\mu = \frac{\delta_{b_3 b_2 b_1 k_1}^{\mu b_2 b_1 k_1}}{\sqrt{\Delta(b_2, b_1, k_1)\Delta(b_3, b_2, b_1, k_1)}} = \frac{i\, \varepsilon^{\mu b_2 b_1 k_1}}{\sqrt{\Delta(b_2, b_1, k_1)}}. \tag{2.47}$$

Like in the triangle case, the $(D - R + 1)$th formally chosen vector b_3 drops out.

2.2.2. Extension of the basis vectors to five space-time dimensions

In the FDH scheme all external momenta are purely four-dimensional, only the loop momentum has additional degrees of freedom to incorporate the dependence on the dimensional regularisation parameter ϵ. A general loop momentum ℓ^μ in $D = 4 - 2\epsilon$ dimensions can therefore be decomposed in terms of a four-dimensional part $\ell_{[4]}^\mu$ and an ϵ-dependent part $\ell_{[-2\epsilon]}^\mu$ which we define to be orthogonal to each other

$$\ell^\mu = \ell_{[4]}^\mu + \ell_{[-2\epsilon]}^\mu \tag{2.48}$$

with the orthogonality condition

$$\ell_{[4]} \cdot \ell_{[-2\epsilon]} = 0. \tag{2.49}$$

Since $\ell_{[-2\epsilon]}^{\mu}$ is orthogonal to the external outflow momenta and to all basis vectors that we have constructed in the previous section, we can think of this vector as an additional transverse vector. In order to have a numerical representation of such a vector, we embed all quantities of the four-dimensional Minkowski space in a five-dimensional space-time and confine the ϵ-degrees of freedom exclusively to the fifth dimension. With respect to the external kinematics, the fifth dimension is thus an entirely transverse subspace. We define the metric in the five-dimensional Minkowski space in analogy to the four-dimensional case as

$$g^{\mu\nu} = \mathrm{diag}(1, -1, -1, -1, -1). \tag{2.50}$$

A suitable choice for a basis vector is then

$$n_{\epsilon}^{\mu} = (0, 0, 0, 0, i) \tag{2.51}$$

which automatically satisfies $n_{\epsilon} \cdot n_{\epsilon} = 1$ and $n_{\epsilon} \cdot k_i = n_{\epsilon} \cdot v_i = n_{\epsilon} \cdot n_i = 0$ where k_i, v_i and n_i are the outflow momenta and the physical respectively the transverse basis vectors in four dimensions. The decomposition of the metric tensor in five dimensions, therefore, reads

$$g^{\mu\nu} = \sum_{i=1}^{D_P} k_i^{\mu} v_i^{\nu} + \sum_{i=1}^{D_T} n_i^{\mu} n_i^{\nu} + n_{\epsilon}^{\mu} n_{\epsilon}^{\nu} \tag{2.52}$$

with the convention that D_P and D_T refer always to the dimension of the external momenta as stated in Tab. 2.1 and the dependence on n_{ϵ} is counted separately.

2.2.3. The loop momentum in terms of the van Neerven-Vermaseren basis

It is now straightforward to express a general loop momentum in terms of the basis vectors v_i, n_i and n_{ϵ} using the decomposition of the metric tensor in Eq. (2.52) [72, 75]:

$$\ell^{\mu} = \delta_{\nu}^{\mu} \ell^{\nu} = \sum_{r=1}^{D_P} (\ell \cdot k_r) v_r^{\mu} + \sum_{i=1}^{D_T} (\ell \cdot n_i) n_i^{\mu} + (\ell \cdot n_{\epsilon}) n_{\epsilon}^{\mu}. \tag{2.53}$$

Next, we express $\ell \cdot k_r$ through inverse propagators and invariants from external momentum sums. To do that, we subtract two subsequent propagators

$$d_{i_{r+1}} = (\ell - q_{r+1})^2 - m_{i_{r+1}}^2$$
$$d_{i_r} = (\ell - q_r)^2 - m_{i_r}^2$$

from each other and use $q_{r+1} - q_r = k_r$ from Eq. (2.4) to get

$$\ell \cdot k_r = -\frac{1}{2} \left[(d_{i_{r+1}} - d_{i_r}) - (q_{r+1}^2 - m_{i_{r+1}}^2) + (q_r^2 - m_{i_r}^2) \right].$$

A general loop momentum therefore reads

$$\ell^\mu = V_R^\mu - \frac{1}{2} \sum_{r=1}^{D_P} (d_{i_{r+1}} - d_{i_r}) v_r^\mu + \sum_{i=1}^{D_T} (\ell \cdot n_i) \, n_i^\mu + (\ell \cdot n_\epsilon) \, n_\epsilon^\mu \qquad (2.54)$$

where we have defined

$$V_R^\mu = +\frac{1}{2} \sum_{r=1}^{D_P} \left[(q_{r+1}^2 - m_{i_{r+1}}^2) - (q_r^2 - m_{i_r}^2) \right] v_r^\mu. \qquad (2.55)$$

For vanishing propagator masses $m_{i_r} = 0$ or equal propagator masses $m_{i_r} = m$ where $r = 1, \ldots, R$, the following identity holds:

$$V_R \cdot q_r = \frac{1}{2} q_r^2. \qquad (2.56)$$

Comparing Eq. (2.54) with Eq. (2.48), it is immediately clear that we can identify the ϵ-dependent part of the loop momentum with $\ell_{[-2\epsilon]}^\mu = (\ell \cdot n_\epsilon) \, n_\epsilon^\mu$ and the four-dimensional part $\ell_{[4]}^\mu$ with the remaining terms.

2.3. The general structure of the one-loop integrand

We now introduce a specific representation of the integrand which is also known as the OPP basis referring to the work of Ossola, Papadopoulos and Pittau [68]. We write the one-loop n-point amplitude in terms of an integral over a single fraction

$$A_n^{\text{1-loop}} = \int \frac{\mathrm{d}^D \ell}{i\pi^{D/2}} \, \frac{\mathcal{N}(\{p_i\}, \{J_i\}, \ell)}{d_1 d_2 \ldots d_n}. \qquad (2.57)$$

The denominator is the product of all possible inverse propagators d_i whereas the numerator $\mathcal{N}(\{p_i\}, \{J_i\}, \ell)$ is a function which depends on the external momenta p_i, the sources of the external particles J_i (polarisation vectors and spinors), masses and on the loop momentum ℓ. This is nothing else than writing the sum of all Feynman diagrams in terms of the least common denominator. In a renormalisable quantum field theory, the power r of the loop momentum $\ell^{\mu_1} \ldots \ell^{\mu_r}$ in a Feynman integral — the tensor rank — may not exceed the multiplicity m of the integral. Hence, with $m \leq n$, at most m-point tensor integrals of rank m occur. One can show that in D integer space-time dimensions, every m-point tensor integral of rank r where $r \leq m$ can be reduced to a D-point tensor integral of rank $r \leq D$ [17]. Exploiting this result and taking into account that we are interested in the case $D = 5$ (four physical space-time dimensions and one extra dimension to incorporate the dimensional regulator), the most general form of the integrand $\mathcal{G}_n$ is a sum of rational functions with five, four, three, two or one inverse propagator(s) in the denominator

$$\mathcal{G}_n(\ell) \equiv \frac{\mathcal{N}(\{p_i\}, \{J_i\}, \ell)}{d_1 d_2 \dots d_n} = \sum_{[i_1|i_5]} \frac{\bar{e}_{i_1 i_2 i_3 i_4 i_5}(\ell)}{d_{i_1} d_{i_2} d_{i_3} d_{i_4} d_{i_5}} + \sum_{[i_1|i_4]} \frac{\bar{d}_{i_1 i_2 i_3 i_4}(\ell)}{d_{i_1} d_{i_2} d_{i_3} d_{i_4}}$$
$$+ \sum_{[i_1|i_3]} \frac{\bar{c}_{i_1 i_2 i_3}(\ell)}{d_{i_1} d_{i_2} d_{i_3}} + \sum_{[i_1|i_2]} \frac{\bar{b}_{i_1 i_2}(\ell)}{d_{i_1} d_{i_2}} + \sum_{[i_1|i_1]} \frac{\bar{a}_{i_1}(\ell)}{d_{i_1}} . \quad (2.58)$$

We refer to the numerator functions $\bar{e}(\ell)$, $\bar{d}(\ell)$, $\bar{c}(\ell)$, $\bar{b}(\ell)$ and $\bar{a}(\ell)$ as pentagon, box, triangle, bubble and tadpole contribution. A complete description of the tensor structure of the numerator functions $\bar{e}$, $\bar{d}$, $\bar{c}$, $\bar{b}$ and $\bar{a}$ in terms of the NV-basis is given Ref. [70]. We will now derive the general functional form of the pentagon and box contribution repeating the arguments that were given in Ref. [75]. This illustrates the concept which applies in an analogous way also to triangle, bubble and tadpole contributions.

The tensor structure of the pentagon contribution: From the above considerations we know that the maximal tensor rank of the five-point function is five, hence, a distinct numerator function has the functional form

$$\mathcal{N}_5(\ell) = W_{\mu_1 \mu_2 \mu_3 \mu_4 \mu_5} \ell^{\mu_1} \ell^{\mu_2} \ell^{\mu_3} \ell^{\mu_4} \ell^{\mu_5} = \prod_{j=1}^{5} (u_j \cdot \ell) . \quad (2.59)$$

where u_j is a combination of external, purely four-dimensional vectors.[2] Using Eq. (2.54), the loop-momentum in terms of the NV-basis reads

$$\ell^\mu = V_5^\mu - \frac{1}{2} \sum_{r=1}^{4} (d_{i_{r+1}} - d_{i_r}) v_r^\mu + (\ell \cdot n_\epsilon) n_\epsilon^\mu . \quad (2.60)$$

We exploit the fact that the u_j are purely four-dimensional, hence $u_j \cdot n_\epsilon = 0$. The functional form of $u_j \cdot \ell$ is therefore a constant term which involves only external momenta plus terms which depend linearly on inverse propagators

$$u_j \cdot \ell = (u_j \cdot V_5) - \frac{1}{2} \sum_{r=1}^{4} (d_{i_{r+1}} - d_{i_r}) (v_r \cdot u_j) . \quad (2.61)$$

Substituting this into Eq. (2.59)

$$\mathcal{N}_5(\ell) = (u_5 \cdot V_5) \prod_{j=1}^{4} (u_j \cdot \ell) - \frac{1}{2} \sum_{r=1}^{4} (d_{i_{r+1}} - d_{i_r}) (v_r \cdot u_5) \prod_{j=1}^{4} (u_j \cdot \ell) \quad (2.62)$$

---

[2]The total pentagon numerator does in general not factorise and is expressed as $\sum_k \prod_{j=1}^{5} u_j^k \cdot \ell$ with u_j^k a combination of external vectors. Since the tensor reduction applies for every single summand, the structure in Eq. (2.59) is sufficient for the following arguments.

and cancelling the inverse propagators, we have reduced the integrand to a rank four five-point tensor integral and a rank four four-point integral. We can repeat the same procedure with the rank four five-point integral to reduce it to a rank three five-point and a rank three four-point integral. Finally we end up with a scalar five-point integral and four-point integrals from rank four down to rank zero. Hence, we conclude that the numerator function of the pentagon contribution is constant, i.e. it does not depend on the loop momentum

$$\bar{e}_{i_1 i_2 i_3 i_4 i_5}(\ell) = e^{(0)}_{i_1 i_2 i_3 i_4 i_5}. \tag{2.63}$$

We note at this point that a constant $e^{(0)}_{i_1 i_2 i_3 i_4 i_5}$ depends on the choice of the master integrals. An alternative description of a non-constant pentagon numerator is given in Ref. [131].

The tensor structure of the box contribution: We apply again the master formula Eq. (2.54) to decompose the loop momentum as

$$\ell^\mu = V_4^\mu - \frac{1}{2}\sum_{r=1}^{3}(d_{i_{r+1}} - d_{i_r})\, v_r^\mu + (\ell \cdot n_1)\, n_1^\mu + (\ell \cdot n_\epsilon)\, n_\epsilon^\mu. \tag{2.64}$$

We start with the highest induced box contribution that we found in the previous pentagon analysis, the rank four four-point function. We simply repeat the arguments from the previous paragraph and get

$$\mathcal{N}_4(\ell) = (u_4 \cdot \ell)\left(\prod_{j=1}^{3} u_j \cdot \ell\right) = (u_4 \cdot V_4)\left(\prod_{j=1}^{3} u_j \cdot \ell\right)$$

$$- \frac{1}{2}\sum_{r=1}^{3}(d_{i_{r+1}} - d_{i_r})\,(u_4 \cdot v_r)\left(\prod_{j=1}^{3} u_j \cdot \ell\right) + (u_4 \cdot n_1)(\ell \cdot n_1)\left(\prod_{j=1}^{3} u_j \cdot \ell\right).$$

The first term is reduced to a rank three four-point function and the second one to a rank three three-point function, both structures are thus of lower tensor rank. The third term, however, is still a rank four tensor. Note that although the u_j vectors are constructed out of external momenta, $u_4 \cdot n_1$ does not necessarily vanish. We apply the same procedure on the last term once again. This will produce reduced rank three three-point and four-point tensors of the form $(\prod_{j=1}^{2} u_j \cdot \ell)(\ell \cdot n_1)$ and a rank four four-point tensor of the form

$$\left(\prod_{j=1}^{2} u_j \cdot \ell\right)(\ell \cdot n_1)^2.$$

Due to the orthogonality of the transverse vectors, we obtain an expression for $(\ell \cdot n_1)^2$ by squaring the loop momentum ℓ in Eq. (2.64) and using $\ell^2 = d_{i_2} + m_{i_2}^2$

$$(\ell \cdot n_1)^2 = -(\ell \cdot n_\epsilon)^2 - V_4^2 + m_{i_2}^2 + d_{i_2} + \sum_{r=1}^{3} (d_{i_{r+1}} - d_{i_r})(v_r \cdot V_4)$$

$$- \frac{1}{4} \sum_{r,s=1}^{3} (d_{i_{r+1}} - d_{i_r})(d_{i_{s+1}} - d_{i_s})(v_r \cdot v_s). \tag{2.65}$$

Inserting this expression in the remaining rank four tensor structure $(\prod_{j=1}^{2} u_j \cdot \ell)(\ell \cdot n_1)^2$, the constant terms will lead to rank two four-point tensors and the terms with propagators to lower point rank two tensors. The only surviving rank four structures are tensors in the $[-2\epsilon]$-components of the loop momentum. We can repeat this procedure now also with the remaining $\prod_j u_j \cdot \ell$ and arrive at the general tensor structure of the box contribution

$$\bar{d}_{i_1 i_2 i_3 i_4}(\ell) = \tilde{d}^{(0)} + \tilde{d}^{(1)}(\ell \cdot n_1) + \tilde{d}^{(2)}(\ell \cdot n_\epsilon)^2 + \tilde{d}^{(3)}(\ell \cdot n_\epsilon)^2(\ell \cdot n_1) + \tilde{d}^{(4)}(\ell \cdot n_\epsilon)^4. \tag{2.66}$$

We suppressed the subscripts $i_1 i_2 i_3 i_4$ at any coefficient $\tilde{d}^{(i)}$ in order to keep things better readable. We stress that the coefficients $\tilde{d}^{(i)}$ do not depend on the loop momentum. They are functions of purely four-dimensional external quantities and particle masses. There are five independent tensor structures from rank zero up to rank four. The terms that are proportional to $(\ell \cdot n_1)$ are so called *spurious terms* which vanish after integration over the loop momentum. To see why this is true, for example for $\tilde{d}^{(1)}(\ell \cdot n_1)$, we recall from the Passarino-Veltman reduction [16] that a rank one four-point tensor integral with the three independent outflow momenta k_1, k_2 and k_3 can be written as

$$\int \frac{\mathrm{d}^D \ell}{(2\pi)^D} \frac{\ell^\mu}{d_1 d_2 d_3 d_4} = k_1^\mu D_{11} + k_2^\mu D_{12} + k_3^\mu D_{13}. \tag{2.67}$$

D_{11}, D_{12} and D_{13} are linear combinations of scalar four-point and scalar three-point integrals. The spurious term will therefore lead to contributions proportional to $(k_i \cdot n_1)$, but since n_1 lives in the transverse space, these contributions vanish after integration over the loop momentum. Also the "mixed" contribution $\tilde{d}^{(3)}(\ell \cdot n_\epsilon)^2(\ell \cdot n_1)$ vanishes after integration over the four-dimensional part of the loop momentum if we split the integral measure according to $\mathrm{d}^D \ell = \mathrm{d}^4 \ell_{[4]} \, \mathrm{d}^{-2\epsilon} \ell_{[-2\epsilon]}$. More formally, we write

$$\int \frac{\mathrm{d}^D \ell}{(2\pi)^D} \frac{\bar{d}(\ell)}{d_1 d_2 d_3 d_4} = \int \frac{\mathrm{d}^4 \ell}{(2\pi)^4} \frac{\tilde{d}^{(0)} + \tilde{d}^{(1)}(\ell \cdot n_1)}{d_1 d_2 d_3 d_4} + \mathcal{R} = \tilde{d}^{(0)} \int \frac{\mathrm{d}^4 \ell}{(2\pi)^4} \frac{1}{d_1 d_2 d_3 d_4} + \mathcal{R}.$$

In $\mathcal{R}$ we have collected the finite rational contributions from the ϵ-dependent part of the loop integration which we will treat separately in section 2.4. Hence, $\tilde{d}^{(0)}$ is precisely the desired four-dimensional integral coefficient. This leads directly to the OPP idea: Assuming that we manage to isolate a particular box integrand $\bar{d}_{i_1 i_2 i_3 i_4}(\ell)$, we can simply evaluate the integrand for five different loop momenta. This generates a system of equations that we can solve for each phase space point to extract the coefficients $\tilde{d}^{(i)}$. We will see in this section that the integrand can be generated from products of tree level amplitudes.

Another important observation is that $(\ell \cdot n_\epsilon)$ appears only in even powers. This follows immediately from Eq. (2.65), or, equivalently, from the structure of our basis vectors in five dimensional Minkowski space: Since all external vectors are four-dimensional and, hence, orthogonal to n_ϵ, the only way how n_ϵ components can survive are contractions with itself. Let us assume that the fifth component of the loop-momentum is μ. With the definition of the five-dimensional metric in Eq. (2.50) and the vector n_ϵ^ν in Eq. (2.51), it is clear that

$$(\ell \cdot n_\epsilon)^2 = -\mu^2 \,. \tag{2.68}$$

For reasons of compactness, we express from now on quantities in terms of μ^2 instead of $(\ell \cdot n_\epsilon)^2$. We further define the shorthand notation

$$(\ell \cdot n_i) = \alpha_i \,. \tag{2.69}$$

Redefining $d^{(0)} \equiv \tilde{d}^{(0)}$, $d^{(1)} \equiv \tilde{d}^{(1)}$, $d^{(2)} \equiv -\tilde{d}^{(2)}$, $d^{(3)} \equiv -\tilde{d}^{(3)}$ and $d^{(4)} \equiv \tilde{d}^{(4)}$, the tensor structure of the box integrand reads

$$\bar{d}_{i_1 i_2 i_3 i_4}(\ell) = d^{(0)} + d^{(1)}\alpha_1 + d^{(2)}\mu^2 + d^{(3)}\mu^2\alpha_1 + d^{(4)}\mu^4. \tag{2.70}$$

We recover the four-dimensional case with its two independent tensor structures setting $\mu = 0$.

The tensor structure of the triangle contribution: A similar analysis as in the box case can be applied to the triangle case. We do not repeat the derivation and simply quote the parametrisation of the triangle tensor structure from Ref. [70]. Using again Eqs. (2.68) and (2.69), $\bar{c}_{i_1 i_2 i_3}(\ell)$ reads

$$\begin{aligned}
\bar{c}_{i_1 i_2 i_3}(\ell) = {}& c^{(0)} + c^{(1)}\,\alpha_1 + c^{(2)}\,\alpha_2 + c^{(3)}\,(\alpha_1^2 - \alpha_2^2) + c^{(4)}\,\alpha_1\alpha_2 \\
& + c^{(5)}\,\alpha_1^2\alpha_2 + c^{(6)}\,\alpha_1\alpha_2^2 + c^{(7)}\,\alpha_1\mu^2 + c^{(8)}\,\alpha_2\mu^2 + c^{(9)}\,\mu^2
\end{aligned} \tag{2.71}$$

where we have suppressed the indices $i_1 i_2 i_3$ at any $c^{(i)}$. $\bar{c}_{i_1 i_2 i_3}(\ell)$ is a polynomial in α_1, α_2 and μ with ten independent structures whose maximal power is three. Similar arguments as in the box case can be applied to show that the terms $c^{(1)}, \ldots, c^{(8)}$ involving different monomials of α_i vanish after integration, i.e. they are all spurious terms. The non-vanishing terms are either the four-dimensional triangle integral coefficient $c^{(0)}$ or a contribution to the rational part from $c^{(9)}$. We recover the seven independent tensor structures of the cut-constructible part $c^{(0)}, \ldots, c^{(6)}$ setting again $\mu = 0$.

The tensor structure of the bubble contribution: Using again Eqs. (2.69) and (2.68) the independent tensor structures of $\bar{b}_{i_1 i_2}(\ell)$ can be parametrised like in Ref. [70] as

$$\bar{b}_{i_1 i_2}(\ell) = b^{(0)} + b^{(1)}\,\alpha_1 + b^{(2)}\,\alpha_2 + b^{(3)}\,\alpha_3 + b^{(4)}\,(\alpha_1^2 - \alpha_3^2) + b^{(5)}\,(\alpha_2^2 - \alpha_3^2)$$
$$+ b^{(6)}\alpha_1\alpha_2 + b^{(7)}\alpha_1\alpha_3 + b^{(8)}\alpha_2\alpha_3 + b^{(9)}\,\mu^2 \tag{2.72}$$

where we have suppressed the indices $i_1 i_2$ at any $b^{(i)}$. There are now four variables α_1, α_2, α_3 and μ contributing to the polynomial with ten independent structures whose maximal power is two. The terms $b^{(1)}, \ldots, b^{(8)}$ involving different monomials of α_i are all spurious and vanish after integration over the loop momentum. The non-vanishing terms are either the four-dimensional bubble integral coefficient $b^{(0)}$ or a contribution to the rational part from $b^{(9)}$. We recover the nine independent tensor structures of the cut-constructible part $b^{(0)}, \ldots, b^{(8)}$ setting $\mu = 0$.

Projecting out the numerators: We now explain, how to project out the numerator functions $\bar{e}(\ell)$, $\bar{d}(\ell)$, $\bar{c}(\ell)$, $\bar{b}(\ell)$ and $\bar{a}(\ell)$ of a distinct topology $C_R = \{i_1, \ldots, i_R\}$ in the OPP basis Eq. (2.58). We start with the pentagon contribution $\bar{e}_{i_1 i_2 i_3 i_4 i_5}(\ell)$. First, we multiply both sides of Eq. (2.58) with the inverse propagators $d_{i_1} d_{i_2} d_{i_3} d_{i_4} d_{i_5}$. If we choose the loop momentum on-shell such that all inverse propagators d_{i_r} of this distinct topology C_R vanish, then the only remaining term of the expansion on the right hand side is the term where all propagators cancel. This is precisely $\bar{e}_{i_1 i_2 i_3 i_4 i_5}(\ell)$. Terms with uncancelled propagators will be nullified as soon as the loop momentum goes on-shell. This is necessarily true for all other pentagons and for all topologies with less than five inverse propagators in the denominator — boxes, triangles, bubbles and tadpoles. On the other hand, the left hand side of Eq. (2.58) multiplied with $d_{i_1} \ldots d_{i_5}$ factorises for on-shell loop momenta into products of five *tree-level amplitudes*. In order to see the factorisation explicitly, we work in axial light-cone gauge, a physical gauge in which no ghosts appear, c.f. for example Ref. [132]. The gluon and quark–gluon vertices as well as the quark propagator are the same as in Lorentz Feynman gauge given in chapter 1. The essential difference is the gluon propagator reading

$$\mu \;\substack{\text{OOO}}\; \nu \;\; = \frac{i}{k^2 + i\varepsilon}\left(-g_{\mu\nu} + \frac{k_\mu q_\nu + q_\mu k_\nu}{k \cdot q}\right) \tag{2.73}$$

with q_μ a light-like reference momentum to fix the gauge. If the loop momentum is on the mass-shell, the numerator of the corresponding propagator (more precisely the factor that multiplies i/d_{i_r}) can be replaced by polarisation sums, exploiting the completeness relation for external wave functions. For example, for massless vector bosons with on-shell momentum ℓ this reads

$$\sum_{\lambda=1}^{D_s-2} \epsilon_\mu^{(\lambda)} \epsilon_\nu^{(\lambda)} = -g_{\mu\nu} + \frac{\ell_\mu q_\nu + q_\mu \ell_\nu}{\ell \cdot q}. \tag{2.74}$$

in agreement with Eq. (2.73). The sum is over the polarisation degrees of freedom where D_s denotes the scheme-dependent spin dimension within the loop with $D_s \geq D$ [70]. Exploiting the completeness relation, thus, cuts a loop propagator into two pieces. Attaching the on-shell wave functions of the cut propagator to the remaining parts of the diagram, one ends up with tree-level structures. Hence, cutting R loop propagators in the sum of all one-loop diagrams leads to a product of R on-shell tree-level amplitudes which needs to be summed over the internal spin degrees of freedom from the R polarisation sums. With five on-shell propagators, the projected pentagon function reads

$$\bar{e}_{i_1 i_2 i_3 i_4 i_5}(\ell) = \sum_{h_1 \ldots h_5} A(-\ell_{i_1}^{-h_1}, p_{i_1}, \ldots, p_{i_2-1}, \ell_{i_2}^{h_2}) \, A(-\ell_{i_2}^{-h_2}, p_{i_2}, \ldots, p_{i_3-1}, \ell_{i_3}^{h_3}) \times$$
$$A(-\ell_{i_3}^{-h_3}, p_{i_3}, \ldots, p_{i_4-1}, \ell_{i_4}^{h_4}) \, A(-\ell_{i_4}^{-h_4}, p_{i_4}, \ldots, p_{i_5-1}, \ell_{i_5}^{h_5}) \times$$
$$A(-\ell_{i_5}^{-h_5}, p_{i_5}, \ldots, p_{i_1-1}, \ell_{i_1}^{h_1}). \tag{2.75}$$

Working in the FDH scheme where $D_s = 4$, our convention is to connect tree-level amplitudes of opposite helicity and momentum at any cut propagator (h_i denotes the helicity of both quarks and gluons). An explicit representation of the polarisation vectors and spinors that satisfy the corresponding completeness relations is given in appendix A. Since everything is on-shell, the tree-level amplitudes are gauge invariant. Hence we are not restricted to axial gauge for the computation of the tree-level amplitudes, instead we can use the tree-level techniques with Lorentz Feynman gauge from chapter 1.

The procedure outlined above can be seen as taking the residue of the integrand, more generally for an arbitrary function $\mathfrak{G}(\ell)$ defined as

$$\text{Res}_{[i_1 i_2 \ldots i_R]}\left(\mathfrak{G}(\ell)\right) = \left(d_{i_1}(\ell) \, d_{i_2}(\ell) \ldots d_{i_R}(\ell) \, \mathfrak{G}(\ell)\right)\Big|_{d_{i_1} = \ldots = d_{i_R} = 0}. \tag{2.76}$$

If $\mathfrak{G}(\ell)$ is equal to the integrand $\mathcal{G}_n(\ell)$, the residue is the product of tree-level amplitudes according to

$$\text{Res}_{[i_1 i_2 \ldots i_R]}\left(\mathcal{G}_n(\ell)\right) = \sum_{h_1 \ldots h_R} A(-\ell_{i_1}^{-h_1}, p_{i_1}, \ldots, p_{i_2-1}, \ell_{i_2}^{h_2}) \, A(-\ell_{i_2}^{-h_2}, p_{i_2}, \ldots, p_{i_3-1}, \ell_{i_3}^{h_3})$$
$$\times \ldots \times A(-\ell_{i_R}^{-h_R}, p_{i_R}, \ldots, p_{i_1-1}, \ell_{i_1}^{h_1}). \tag{2.77}$$

The topology C_R selects R loop momentum flavours from the parent diagram's propagator list $\mathcal{F}_{\text{par}} = [F_1, \ldots, F_N]$ to assign the flavours of the tree-level amplitudes. In case that they involve a dummy propagator, the whole residue is set equal to zero. For the correct projection onto the box, triangle and bubble numerators higher topologies have to be subtracted. It follows from Eq. (2.58) that multiplying with $R < M \leq 5$ inverse propagators, there are contributions from topologies with M propagators which have the R propagators in common and will entirely cancel. Hence, besides the desired numerator function, higher topology structures with common propagators will not vanish and must be explicitly subtracted. We arrive, thus, at the following expressions for the pentagon, box, triangle and bubble numerators [70]:

$$\bar{e}_{i_1i_2i_3i_4i_5}(\ell) = \operatorname{Res}_{[i_1i_2i_3i_4i_5]}\Big(\mathcal{G}_n(\ell)\Big) \tag{2.78}$$

$$\bar{d}_{i_1i_2i_3i_4}(\ell) = \operatorname{Res}_{[i_1i_2i_3i_4]}\left(\mathcal{G}_n(\ell) - \sum_{[i_1|i_5]}\frac{\bar{e}_{i_1i_2i_3i_4i_5}(\ell)}{d_{i_1}d_{i_2}d_{i_3}d_{i_4}d_{i_5}}\right) \tag{2.79}$$

$$\bar{c}_{i_1i_2i_3}(\ell) = \operatorname{Res}_{[i_1i_2i_3]}\left(\mathcal{G}_n(\ell) - \sum_{[i_1|i_5]}\frac{\bar{e}_{i_1i_2i_3i_4i_5}(\ell)}{d_{i_1}d_{i_2}d_{i_3}d_{i_4}d_{i_5}} - \sum_{[i_1|i_4]}\frac{\bar{d}_{i_1i_2i_3}(\ell)}{d_{i_1}d_{i_2}d_{i_3}d_{i_4}}\right) \tag{2.80}$$

$$\bar{b}_{i_1i_2}(\ell) = \operatorname{Res}_{[i_1i_2]}\left(\mathcal{G}_n(\ell) - \sum_{[i_1|i_5]}\frac{\bar{e}_{i_1i_2i_3i_4i_5}(\ell)}{d_{i_1}d_{i_2}d_{i_3}d_{i_4}d_{i_5}} - \sum_{[i_1|i_4]}\frac{\bar{d}_{i_1i_2i_3}(\ell)}{d_{i_1}d_{i_2}d_{i_3}d_{i_4}}\right.$$
$$\left. - \sum_{[i_1|i_3]}\frac{\bar{c}_{i_1i_2i_3}(\ell)}{d_{i_1}d_{i_2}d_{i_3}}\right) \tag{2.81}$$

In order to illustrate this point we give an explicit example of the five-point triangle residue $\bar{c}_{134}(\ell)$

$$\bar{c}_{134}(\ell) = \operatorname{Res}_{[123]}\Big(\mathcal{G}_n(\ell)\Big) - \frac{\bar{d}(\ell)_{1234}}{d_2} - \frac{\bar{d}(\ell)_{1345}}{d_5} - \frac{\bar{e}(\ell)_{12345}}{d_2d_5}.$$

We summarise once again the basic ideas of this section: Setting different sets of propagators on the mass-shell allows to project out the numerator functions of the integrand in Eq. (2.57). Since going on-shell factorises the integrand in terms of tree-level amplitudes, one can construct the numerator functions directly taking products of tree-level amplitudes. Tree-level techniques are by now very well developed, hence, the computation of the numerator functions is straightforward. Evaluating the integrand for as many independent on-shell momenta as there are independent tensor structures, generates a system of equations that can be solved to disentangle both the integral coefficients and the spurious terms from each other. We will describe in the next sections how this is done in practice.

2.4. Rational part versus cut-constructible part and the choice of master integrals

From now on, we restrict our arguments to strictly massless QCD. Following the arguments in section 2.3 that all spurious terms vanish after integration, we are left with scalar 5-point, 4-point, 3-point and 2-point master integrals multiplying the four-dimensional integral coefficients plus additional master integrals which depend on the transverse part of the loop momentum μ^2. One can read off from Eqs. (2.70), (2.71) and (2.72) that these additional master integrals are rank four and rank two tensor box integrals, rank two tensor triangle integrals and rank two tensor bubble integrals. Note, however, that the tensor structure appears exclusively in μ^2, i.e. only in the epsilon dependent part of the loop momentum. Following the arguments given in Ref. [70], one can rewrite these master integrals in higher dimensional space-time and then take the limit $\epsilon \to 0$. The finite remainder is precisely the rational part. One finds for the additional master integrals [70]

$$\int \frac{\mathrm{d}^D \ell}{i\,\pi^{D/2}} \frac{\mu^2}{d_{i_1} d_{i_2} d_{i_3} d_{i_4}} = -\frac{D-4}{2} \mathcal{I}^{[D+2]}_{4;i_1 i_2 i_3 i_4},$$

$$\int \frac{\mathrm{d}^D \ell}{i\,\pi^{D/2}} \frac{\mu^4}{d_{i_1} d_{i_2} d_{i_3} d_{i_4}} = +\frac{(D-4)(D-2)}{4} \mathcal{I}^{[D+4]}_{4;i_1 i_2 i_3 i_4},$$

$$\int \frac{\mathrm{d}^D \ell}{i\,\pi^{D/2}} \frac{\mu^2}{d_{i_1} d_{i_2} d_{i_3}} = -\frac{D-4}{2} \mathcal{I}^{[D+2]}_{3;i_1 i_2 i_3},$$

$$\int \frac{\mathrm{d}^D \ell}{i\,\pi^{D/2}} \frac{\mu^2}{d_{i_1} d_{i_2}} = -\frac{D-4}{2} \mathcal{I}^{[D+2]}_{2;i_1 i_2}.$$

The six-dimensional scalar box integral is UV finite, hence multiplying with $(D-4) = -2\epsilon$ vanishes in the four-dimensional limit. However, the eight-dimensional scalar box integral, the six-dimensional scalar triangle integral and the six-dimensional scalar bubble integral are UV divergent and lead to ϵ^{-1}-poles. Multiplying with $(D-4)$ gives thus finite contributions in the four-dimensional limit [70]:

$$\lim_{D \to 4} \frac{(D-4)(D-2)}{4} \mathcal{I}^{[D+4]}_{4;i_1 i_2 i_3 i_4} = -\frac{1}{6}$$

$$\lim_{D \to 4} \frac{D-4}{2} \mathcal{I}^{[D+2]}_{3;i_1 i_2 i_3} = \frac{1}{2}$$

$$\lim_{D \to 4} \frac{D-4}{2} \mathcal{I}^{[D+2]}_{2;i_1 i_2} = -\frac{1}{6}(q_{i_1} - q_{i_2})^2$$

The rational part can therefore be written as [70]:

$$\mathcal{R} = -\sum_{[i_1|i_4]} \frac{d^{(4)}_{i_1 i_2 i_3 i_4}}{6} - \sum_{[i_1|i_3]} \frac{c^{(9)}_{i_1 i_2 i_3}}{2} + \sum_{[i_1|i_2]} \frac{(q_{i_1} - q_{i_2})^2}{6} b^{(9)}_{i_1 i_2}. \tag{2.82}$$

The structure of a one-loop amplitude as it is introduced in Eq. (2.8) does not involve pentagons. Since any scalar five point function in four dimensions can be written in terms of scalar four-point functions [17, 19], one can indeed express the cut-constructible part in terms of scalar box, triangle and bubble integrals. Hence, the full amplitude can be written as

$$A_n^{\text{1-loop}} = \mathcal{A}^{cc} + \mathcal{R} \tag{2.83}$$

where the rational part $\mathcal{R}$ is defined in Eq. (2.82) and the cut-constructible part $\mathcal{A}^{cc}$

reads

$$\mathcal{A}^{cc} = \sum_{[i_1|i_4]} \tilde{d}^{(0)}_{i_1 i_2 i_3 i_4} \mathcal{I}^{[4-2\epsilon]}_{4;i_1 i_2 i_3 i_4} + \sum_{[i_1|i_3]} c^{(0)}_{i_1 i_2 i_3} \mathcal{I}^{[4-2\epsilon]}_{3;i_1 i_2 i_3} + \sum_{[i_1|i_2]} b^{(0)}_{i_1 i_2} \mathcal{I}^{[4-2\epsilon]}_{2;i_1 i_2}. \tag{2.84}$$

The tilde on $\tilde{d}^{(0)}_{i_1 i_2 i_3 i_4}$ indicates that this coefficient is not necessarily identical to $d^{(0)}_{i_1 i_2 i_3 i_4}$ in Eq. (2.70) if the scalar pentagon integral is considered as a master integral. In our approach, we entirely separate the computation of the cut-constructible part from the rational part, i.e. we do two completely *independent* computations as will be explained below. The cut-constructible part is computed entirely in four space-time dimensions setting $\mu^2 = 0$. In this four-dimensional limit, we are then allowed to ignore the pentagon contribution at the integrand level. An argument to see this has been pointed out in Ref. [131] where the pentagon residue has been parametrised as $\bar{e} = \mu^2 e^{(0)}$ which vanishes for $\mu^2 \to 0$. This is in agreement with the well known fact that the scalar pentagon integral is not independent in four spacetime dimensions as argued above.

For the rational part, the pentagon contributions must be included because at this stage we take explicitly the ϵ-dependence via the fifth component of the loop momentum into account. With the additional degree of freedom, we are able to constrain five propagators on the mass-shell and, hence, to project out the pentagon residue (for whichever employed pentagon parametrisation). Although this contribution is not explicitly present in Eq. (2.82) it enters in the computation of $d^{(4)}_{i_1 i_2 i_3 i_4}$, $c^{(9)}_{i_1 i_2 i_3}$ and $b^{(9)}_{i_1 i_2}$ as subtraction term.

Going from four-dimensional to five-dimensional space-time, however, changes the polarisation degrees of freedom of the particles in the loop. In Ref. [70], the number of spin dimensions D_s and space-time dimensions D are treated as separate integer variables with the constraint $D_s \geq D$. It is further argued that a general one-loop integrand depends linearly on D_s and can be parametrised as

$$\mathcal{N}^{(D,D_s)} = \mathcal{N}_0 + (D_s - 4)\mathcal{N}_1 \tag{2.85}$$

where $\mathcal{N}_0$ and $\mathcal{N}_1$ are independent of D_s. The only way how at one-loop order D_s explicitly enters is from a closed string of Lorentz indices along the loop line. This happens either in diagrams with pure gluonic loop content or in mixed quark–gluon diagrams with at least one gluon propagator in the loop. (For the closed quark loop a D_s-dependence arises at most from the trace normalisation. This is an overall factor which can be taken into account separately at the end of the computation.)

In our approach, we replace for the independent computation of the rational part the gluons inside the loop with real scalars by introducing additional scalar–gluon and scalar–quark vertices. We demonstrate below that this allows the computation of the entire rational part. Since, per definitionem, the scalars do not carry any Lorentz index, the explicit dependence on D_s in Eq. (2.85) vanishes. In addition, this circumvents constructing additional polarisation vectors in higher spin dimensions when constructing the integrand from tree-level amplitudes.

The second important ingredient of the approach is to interpret the additional degrees of freedom from dimensional regularisation as a mass-shift to a four-dimensional loop momentum. Recalling Eq. (2.48) and the orthogonality of the four-dimensional part $\ell^\mu_{[4]}$ with the epsilon dependent part $\ell^\mu_{[-2\epsilon]}$, the mass shifted squared loop momentum reads

$$\ell^2 = \ell^2_{[4]} + \ell^2_{[-2\epsilon]} = \ell^2_{[4]} - \mu^2. \qquad (2.86)$$

If we set a D-dimensional massless loop momentum on the mass shell, i.e. $\ell^2 \overset{!}{=} 0$, we get the on-shell relation of a four-dimensional massive particle:

$$\ell^2_{[4]} = \mu^2. \qquad (2.87)$$

Hence massless unitarity cuts in D space-time dimensions can be considered as four-dimensional massive cuts. Interpreting μ^2 as a uniform mass shift constrained by the various on-shell conditions, we compute the rational part of massless quark and gluon amplitudes by means of amplitudes with massive quarks and massive scalars in the loop. (We comment towards the end of this section how slashing the epsilon dependent part of the loop momentum with gamma matrices works in numerical applications since this is required in fermion propagators.) First ideas to interpret the epsilon dependent part as a mass in the context of unitarity go back to Bern and Morgan described in Ref. [62]. In the context of analytical computations with generalised unitarity, Badger could show that the rational terms can be extracted from the behaviour of the integrand for large (complex) values of μ (remember that μ is still an integration variable) [67]. In the numerical application, μ is kept finite at any stage within the computation. How it is chosen in detail will be explained in section 2.6. The crucial point is that within the presented algorithm, the whole computation is done effectively in four space-time dimensions retaining the conventional spinor helicity formalism. In its present form, it is a priori restricted to massless theories, as will be discussed below.

This is in contrast to the method of D-dimensional unitarity of Ref. [70]: Taking Eq. (2.85) as starting point, the idea is to compute the amplitude twice in different higher *integer* spin dimensions. This generates a system of two equations which can be solved to determine $\mathcal{N}_0$ and $\mathcal{N}_1$. Subsequently, one can interpolate to the physical value $D_s = 4$ for the FDH scheme or to $D_s = 4 - 2\epsilon$ for the 't Hooft-Veltman scheme. This approach requires the explicit construction of polarisation vectors and spinors in higher spin dimensions.

We will derive in the following how to decompose massless QCD amplitudes such that the rational part can be extracted from an integrand where gluons are replaced by real scalar particles. We will use similar arguments as presented in Refs. [55] and [133].

Gluon amplitudes: In Ref. [55] it is shown that gluon amplitudes in QCD can be decomposed in terms of three contributions: An $\mathcal{N} = 4$ supersymmetric amplitude, the contribution of an internal $\mathcal{N} = 1$ chiral matter multiplet (a scalar and a fermionic contribution) and a contribution of complex scalars in the loop. For an n-gluon one-loop amplitude with arbitrary external helicities, the sypersymmetric decomposition reads

$$\mathcal{A}_n^{[g]} = \mathcal{A}_n^{\mathcal{N}=4} - 4\,\mathcal{A}_n^{\mathcal{N}=1\,\text{chiral}} + \mathcal{A}_n^{[s^*]}, \tag{2.88}$$

$$\mathcal{A}_n^{[f]} = \mathcal{A}_n^{\mathcal{N}=1\,\text{chiral}} - \mathcal{A}_n^{[s^*]} \tag{2.89}$$

where $\mathcal{A}_n^{[g]}$ denotes the pure gluon amplitude and $\mathcal{A}_n^{[f]}$ the corresponding amplitudes with a closed quark loop. Since both $\mathcal{A}_n^{\mathcal{N}=4}$ and $\mathcal{A}_n^{\mathcal{N}=1\,\text{chiral}}$ are cut-constructible in four dimensions [55], the rational part of the gluon amplitudes is identical with a gluon amplitude where the gluons circulating in the loop are replaced with complex scalars.

$$\mathcal{A}_n^{[g]}\big|_{\text{rat}} = +\mathcal{A}_n^{[s^*]}\big|_{\text{rat}}, \tag{2.90}$$

$$\mathcal{A}_n^{[f]}\big|_{\text{rat}} = -\mathcal{A}_n^{[s^*]}\big|_{\text{rat}}. \tag{2.91}$$

From this decomposition, one can immediately read off that the rational part of gluon amplitudes with closed quark loop agrees with the rational part of pure gluon amplitudes up to an overall minus sign. We exploit this fact in NJET which leads to a significant simplification of the computation.

We will now show by a diagrammatic power counting argument that one can replace the gluons with real scalars to compute the rational part avoiding arguments from supersymmetry. The results for the pure gluon case can then be generalised to the mixed quark-gluon case. This is of great interest because so far, there is not such a supersymmetric decomposition for amplitudes with arbitrarily many external quark lines like for the pure gluonic case. For the derivation we use background field Feynman gauge [134]. Within this gauge, one can separate in every Feynman diagram the two leading powers of the loop momentum. It turns out that these contributions which spoil the cut-constructibility of the amplitude are equivalent to a scalar contribution.

The basic idea of the background field gauge is to split the gauge field A_μ in a fluctuating quantum field A_μ^Q and a classical background field A_μ^B according to $A_\mu = A_\mu^B + A_\mu^Q$. The functional integral to compute the effective action is then done only with respect to the quantum field A_μ^Q with the gauge fixing condition $D^{B\mu}A_\mu^Q = 0$ and with the background field covariant derivative $D_\mu^B = \partial_\mu - (i/\sqrt{2})gA_\mu^B$. In this way the effective action $\Gamma[A^B]$ which is the generating functional for the one-particle irreducible (1PI) diagrams is a functional of the classical background field A_μ^B. The gauge of A_μ^B is not fixed. The striking feature is that $\Gamma[A^B]$ is independently invariant with respect to gauge transformations of the background field. This implies that any Feynman diagram that contributes to the amplitude can be generated from a 1PI diagram where we can still choose the gauge of the trees sewn to the loop lines. At one-loop order, the computation of a 1PI diagram becomes particularly simple since we can discard all vertices with more than two legs attached to the fluctuating quantum field. From the Feynman rules in Ref. [134] it is straightforward to derive the corresponding colour ordered Feynman rules in background Field Feynman gauge including the ghosts by the

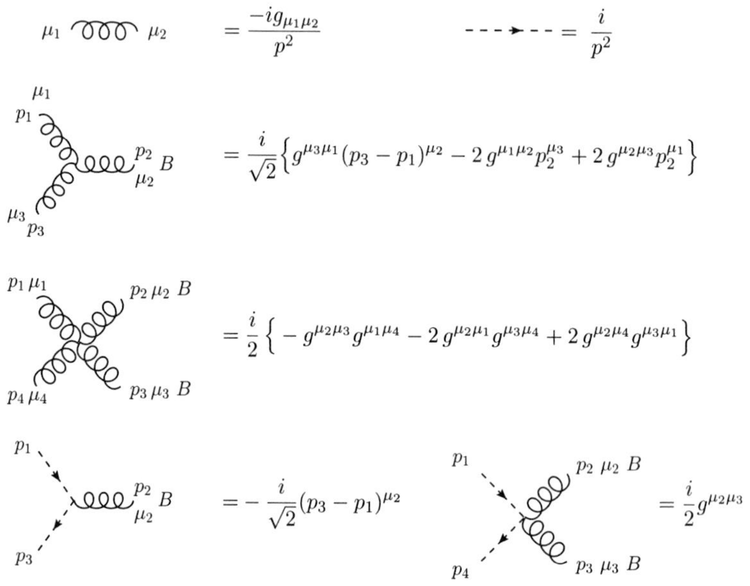
μ_1 μ_2 $= \frac{-ig_{\mu_1\mu_2}}{p^2}$
$= \frac{i}{p^2}$
p_1 μ_1 p_2 μ_2 B μ_3 p_3
$= \frac{i}{\sqrt{2}}\left\{g^{\mu_3\mu_1}(p_3-p_1)^{\mu_2} - 2\,g^{\mu_1\mu_2}p_2^{\mu_3} + 2\,g^{\mu_2\mu_3}p_2^{\mu_1}\right\}$
$p_1\,\mu_1$ $p_2\,\mu_2\,B$ $p_4\,\mu_4$ $p_3\,\mu_3\,B$
$= \frac{i}{2}\left\{-g^{\mu_2\mu_3}g^{\mu_1\mu_4} - 2\,g^{\mu_2\mu_1}g^{\mu_3\mu_4} + 2\,g^{\mu_2\mu_4}g^{\mu_3\mu_1}\right\}$
p_1 p_2 B μ_2 p_3
$= -\frac{i}{\sqrt{2}}(p_3-p_1)^{\mu_2}$
p_1 $p_2\,\mu_2\,B$ p_4 $p_3\,\mu_3\,B$
$= \frac{i}{2}g^{\mu_2\mu_3}$

Figure 2.3.: Colour ordered Feynman rules in background field Feynman gauge. All momenta are taken to be outgoing. The additional B labels the external background field which couples to the internal quantum field.

usual colour stripping methods described in chapter 1. The vertices and propagators are depicted in Fig. 2.3 with all momenta taken to be outgoing. The additional B labels the external background field which couples to the internal quantum field.

In order to investigate the UV behaviour of the loop momentum it is sufficient to investigate all possible 1PI m-point diagrams with $m \leq n$ where n denotes the multiplicity of the one-loop amplitude. In other words, we do not care which kinds of external trees are attached to the loop. Let us discard for a moment the ghost contributions and focus on gluons inside the loop. The maximum tensor rank ℓ^m may only appear in a 1PI diagram that consists exclusively out of three-point vertices. From the vertices in Fig. 2.3, it follows that only the first term of the three-point vertex depends on the loop momentum while the other two terms involve only the external momentum. Choosing $p_1 = \ell_1$, $p_3 = -\ell_1 - p_2$ and remembering that due to current conservation $p_2 \cdot V(p_2) = 0$ with $V^\mu(p_2)$ being an external gluon current of momentum p_2, an m-point 1PI diagram is proportional to $\sim D_s \prod_{i=1}^m (2\ell_i)$ (ℓ_i is the loop momentum of the ith propagator). The dependence on the number of spin dimensions D_s enters due to the closed string of Lorentz indices. Note that in the FDH scheme $D_s = 4$. The second and the third term in the three-gluon vertex are anti-symmetric in μ_1 and μ_3.

Hence, if $(m - 1)$ loop momentum dependent terms of the three-point vertices hit the loop momentum independent terms of the mth three-point vertex, these contributions will mutually cancel. This means that there is no subleading contribution in the loop momentum $\sim \ell^{m-1}$ generated by three-point vertices exclusively. The only way how such contributions can appear is when $(m - 1)$ loop momentum dependent terms of the three-point vertices hit the first term of the four-point vertex giving a contribution $\sim D_s \prod_{i=1}^{m-1}(2l_i)$. Again, the second and the third term of the four-point vertex are anti-symmetric in μ_1 and μ_4, hence, any contractions $\sim \ell^{m-1}$ will mutually cancel. We conclude that the leading and subleading power of the loop momentum which spoil the cut-constructibility of a diagram arise exclusively from the three-point term $\sim g^{\mu_3\mu_1}(p_3 - p_1)^{\mu_2}$ and from the four-point term $\sim g^{\mu_2\mu_3}g^{\mu_1\mu_4}$. These are precisely the terms with metric tensors connecting the loop lines. Since the contraction of these tensors will always lead to a factor of D_s, the leading and subleading loop momentum can be subtracted diagramwise if the gluons in the loop are replaced with real scalars and the gluon vertices with scalar-gluon vertices with the following Feynman rules:

$$- - - - - - - = \frac{i}{p^2} \tag{2.92}$$

$$V_{3;sgs}^{\mu_2}(p_1, p_2, p_3) = -\frac{i}{\sqrt{2}}(p_3 - p_1)^{\mu_2} \tag{2.93}$$

$$V_{4;sggs}^{\mu_2\mu_3}(p_1, p_2, p_3, p_4) = \frac{i}{2}g^{\mu_2\mu_3} \tag{2.94}$$

A single gluonic 1PI m-point diagram $\mathcal{F}_g$ can now be written as

$$\begin{aligned} \mathcal{F}_g &= (\mathcal{F}_g - D_s\mathcal{F}_s) + D_s\mathcal{F}_s \\ &= \mathcal{F}^{\mathrm{cc}} + D_s\mathcal{F}_s. \end{aligned} \tag{2.95}$$

After subtracting the scalar contribution $D_s\mathcal{F}_s$, $\mathcal{F}^{\mathrm{cc}}$ has at most a loop momentum behaviour $\sim \ell^{m-2}$ and therefore the power counting criterion of Eq. (2.9) is applicable, i.e. $\mathcal{F}^{\mathrm{cc}}$ contributes only to the cut-constructible part. Hence, the rational part of this single diagram would simply be

$$\mathcal{F}_g\big|_{\mathrm{rat}} = D_s\mathcal{F}_s\big|_{\mathrm{rat}}. \tag{2.96}$$

To make a statement on the full amplitude, we have to include also the ghosts inside the loop. The scalar-gluon vertices that we have just introduced "by hand" are precisely identical with the ghost-gluon vertices. Apart from the overall minus sign that every ghost loop picks up, the computation is identical. One might worry that reversing the ghost flux in an m-point diagram would give an additional factor of $(-1)^m$. However, one can show that the colour structure of a diagram with reversed ghost agrees with the original one up to a factor of $(-1)^m$ which compensates the sign of the colour ordered vertex. Hence in this special gauge we may simply ignore the arrows at the ghost lines, treat everything as real scalars and multiply the result by a factor of two. Taking these two degrees of freedom of the ghosts into account, the rational part of the full amplitude is given by

$$\mathcal{A}_n^{[g]}\big|_{\text{rat}} = (D_s - 2)\mathcal{A}_n^{[s]}\big|_{\text{rat}}. \tag{2.97}$$

The factor $D_s - 2 = 2$ (FDH scheme) is already implicitly present in the supersymmetric decomposition in Eq. (2.88) because in that case the scalars are complex and have precisely two degrees of freedom.

Mixed quark–gluon amplitudes: A very similar power counting argument as introduced for the pure gluonic case can also be applied to mixed quark-gluon amplitudes. As a result we may compute the rational part of mixed quark-gluon amplitudes by introducing an additional scalar-quark coupling and simply replacing all remaining gluons in the loop with scalars. We will derive this statement now.

We work again in background field Feynman gauge. The colour ordered quark-gluon couplings in this gauge are the same as in Lorentz Feynman gauge, c.f. Eqs. (1.28). The idea is again to investigate the leading powers of the loop momentum diagram by diagram. We distinguish two separate classes of diagrams:

1. No external quark line enters the loop, the loop content is purely gluonic, all external quark lines are attached via gluonic currents to the gluons inside the loop.

2. At least one external quark line enters the loop and contributes with at least one fermionic propagator.

For the first point, we can completely take over the argument from the pure gluonic case in the previous paragraph. The main point there was that the loop momentum behaviour can entirely be investigated with respect to the 1PI diagram which due to the purely gluonic content coincides with those treated in the previous section. Including again the ghost contribution we have the analogue of Eq. (2.97)

$$\mathcal{A}_n^{[g\,\text{exclusive}]}(\{g_i\}, \{q_j, \bar{q}_j\})\big|_{\text{rat}} = (D_s - 2)\mathcal{A}_n^{[s\,\text{exclusive}]}(\{g_i\}, \{q_j, \bar{q}_j\})\big|_{\text{rat}}. \tag{2.98}$$

It is important to stress that $\mathcal{A}_n^{[g\,\text{exclusive}]}(\{g_i\}, \{q_j, \bar{q}_j\})$ is not necessarily a gauge invariant amplitude, it is rather the sum of those diagrams with external quarks and gluons where exclusively ghosts and gluons are inside the loop.

If a quark line enters the m-point 1PI diagram, the maximum tensor rank is automatically reduced by one power with respect to the pure gluon loop and therefore at most ℓ^{m-1}. To see this, assume that the quark line within the loop involves p quark-gluon vertices as shown in Fig. 2.4. Consequently there are $(p-1)$ quark propagators which contribute a loop momentum $\sim \ell^{p-1}$ and at most $(m-p)$ three-gluon vertices leading to a loop momentum behaviour $\sim \ell^{m-p}$. Hence the maximal tensor rank is $(m-1)$. As a consequence, the tensor rank of any m-point 1PI diagram with k different quark lines entering the loop is at most ℓ^{m-k}. This means that all diagrams with more than one

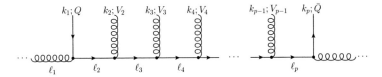

Figure 2.4.: Quark line part of the 1PI m-point diagram.

quark line inside the loop automatically fulfil the power counting criterion and will not contribute to the rational part of the amplitude. To investigate the leading loop momentum which contributes to the rational part, it is therefore sufficient to analyse the m-point 1PI diagrams with a single quark line consisting of $(p-1)$ quark propagators and $(m-p)$ three-gluon vertices. The quark sector of this diagram is shown in Fig. 2.4. The parametrisation of the loop momentum is given by Eqs. 2.5 reading $\ell_1 = \ell + k_1, \ell_2 = \ell, \ell_i = \ell - k_2 \ldots - k_i$. The numerator $\sim \ell^{p-1}$ which this part of the diagram produces has the following functional form

$$
\gamma^\mu \ell_2 \slashed{V}_2 \ell_3 \slashed{V}_3 \ldots \ell_{p-1} \slashed{V}_{p-1} \ell_p \gamma_\mu
$$
$$
= \gamma^\mu \slashed{\ell} \slashed{V}_2 \slashed{\ell} \slashed{V}_3 \ldots \slashed{\ell} \slashed{V}_{p-1} \slashed{\ell} \gamma_\mu + O(\ell^{p-2}) \tag{2.99}
$$

where V_i^μ are vector currents attached to the quark line and we discarded terms of lower power which do not contribute to the rational part. The two contracted Lorentz indices are necessarily equal because each of the $(m-p)$ loop momentum dependent three-gluon vertices provides a metric tensor whose indices connect loop lines. By repeated use of the Dirac algebra $\{\gamma^\mu, \gamma^\nu\} = 2g^{\mu\nu}$ we obtain

$$
\gamma^\mu \slashed{\ell} \gamma_\mu = -(D_s - 2)\, \slashed{\ell}
$$
$$
\gamma^\mu \slashed{\ell} \slashed{V}_2 \slashed{\ell} \gamma_\mu = -(D_s - 2)\, \slashed{\ell} \slashed{V}_2 \slashed{\ell}
$$
$$
\gamma^\mu \slashed{\ell} \slashed{V}_2 \slashed{\ell} \slashed{V}_3 \slashed{\ell} \gamma_\mu = -(D_s - 2)\, \slashed{\ell} \slashed{V}_2 \slashed{\ell} \slashed{V}_3 \slashed{\ell} + \ell^2\, F(\ell^1)
$$
$$
\vdots
$$
$$
\gamma^\mu \slashed{\ell} \slashed{V}_2 \slashed{\ell} \slashed{V}_3 \ldots \slashed{\ell} \slashed{V}_{p-1} \slashed{\ell} \gamma_\mu = -(D_s - 2)\, \slashed{\ell} \slashed{V}_2 \slashed{\ell} \slashed{V}_3 \ldots \slashed{\ell} \slashed{V}_{p-1} \slashed{\ell} + \ell^2\, F(\ell^{p-3}) \tag{2.100}
$$

where $F(\ell^{p-3})$ is a distinct combination of gamma matrices contracted with external kinematics and $(p-3)$ loop momenta. The crucial point is that this function is always multiplied with ℓ^2 which cancels one inverse propagator in the denominator. This reduces the multiplicity of the integral by one unit and the tensor rank by two units leading to an $(m-1)$-point integral of tensor rank $(m-1-2)$ which is cut-constructible. The remaining string of slashed Dirac matrices is the only term which after restoring the loop momenta from the three-gluon vertices will contribute to the rational part. It differs from the original term on the left hand side only by the absence of the unslashed gamma matrices and the presence of the universal factor $-(D_s - 2)$. Hence, the part of a diagram which does not fulfil the power counting criterion, can be factored in a contribution from the closed string of Lorentz indices and a remaining string of gamma matrices contracted with the loop momentum and external quantities leading to the rational part. Yet, the latter contribution can also effectively be computed from a scalar contribution, i.e. we can apply the same trick as in the gluonic case: We add and subtract diagrams in which the gluons are replaced with real scalars (with the scalar–gluon vertices from the previous section) and introduce new quark-scalar vertices instead of the quark–gluon vertices:

$$
\rangle\!\!-\!-\!-\!- = +\frac{i}{\sqrt{2}}\mathbb{1} \qquad\qquad \rangle\!\!-\!-\!-\!- = -\frac{i}{\sqrt{2}}\mathbb{1} \qquad (2.101)
$$

$$
\rangle\!\!-\!-\!-\!- = -\frac{i}{\sqrt{2}}\mathbb{1} \qquad\qquad \rangle\!\!-\!-\!-\!- = +\frac{i}{\sqrt{2}}\mathbb{1} \qquad (2.102)
$$

The double fermion line represents quarks inside the loop, the single fermion line external fermions. The first row of vertices are the left turning ones, the second row the right turning ones. Our convention is such that we multiply a vertex by a factor of (-1) at any time when a quark is leaving the loop. This is necessary to reproduce the minus sign in front of $(D_s - 2)$ in Eqs. (2.100). This is the reason why we distinguish between four different vertices. For an individual diagram with at least one quark propagator in the loop, this reads

$$
\mathcal{F}_{g-q} = (\mathcal{F}_{g-q} - (D_s - 2)\mathcal{F}_{s-q}) + (D_s - 2)\,\mathcal{F}_{s-q}
$$
$$
= \mathcal{F}^{cc} + (D_s - 2)\,\mathcal{F}_{s-q}. \qquad (2.103)
$$

Finally the rational contribution of all diagrams with quarks inside the loop is then simply

$$
\mathcal{A}_n^{[g-q]}(\{g_i\}, \{q_j, \bar{q}_j\})\big|_{\text{rat}} = (D_s - 2)\mathcal{A}_n^{[s-q]}(\{g_i\}, \{q_j, \bar{q}_j\})\big|_{\text{rat}}. \qquad (2.104)
$$

with $\mathcal{A}_n^{[s-q]}$ obtained from $\mathcal{A}_n^{[g-q]}$ by replacing the gluons in the loop with real scalars and applying the above introduced quark–scalar vertices. We stress again that this is so far only a collection of diagrams. To get the full contribution for the rational part we must add also the gluon loop contribution in Eq. 2.98. The rational part of a mixed quark–gluon amplitude with arbitrarily many external quarks is then simply

$$
\mathcal{A}_n(\{g_i\}, \{q_j, \bar{q}_j\})\big|_{\text{rat}} = (D_s - 2)(\mathcal{A}_n^{[s\,\text{exclusive}]} + \mathcal{A}_n^{[s-q]})\big|_{\text{rat}}. \qquad (2.105)
$$

The computation of the rational part of gluon amplitudes and mixed quark-gluon amplitudes is thus equivalent to computing the rational terms of a theory with scalars instead of gluons inside the loop. The additional vertices are given in Eqs. (2.92) – (2.94) and (2.101) – (2.102).

We conclude with a subtlety on the gauge choice. In section 2.3, we argued with axial gauge to show that the integrand factorises into products of tree-level amplitudes since in this gauge the ghosts vanish. One can question whether we are allowed to compute the rational part with a different gauge from the one we computed the cut-constructible part with. In fact, this is not a problem since both parts are separately gauge invariant. Consider first the full gauge invariant amplitude with integral coefficients in D dimensions as given in Eq. (2.7). Since the scalar master integrals form a basis and since the full amplitude is gauge invariant, the D-dimensional integral coefficients are equal for both gauges as long as we reduce the amplitude in both cases to the same set of scalar master integrals. This implies that also the expanded integral coefficients in four dimensions are equal and therefore also necessarily the rational part.

Comments on massive fermion propagators and the fate of μ: Since we evaluate the integrand numerically, we must give an explicit description for the fermion propagators how to slash the five-dimensional loop momentum with gamma matrices $\ell = \ell_{[4]} + \mu\!\!\!/$. (We denote in the following with μ interchangeably either a D-dimensional vector orthogonal to four-dimensional space-time or the fifth component of a five-dimensional vector.) Due to the even nature of the Dirac algebra, a general and consistent description would require to construct a representation of the Dirac algebra in at least $D_s = 6$ dimensions giving rise to six 8×8 gamma matrices [71]. We show that in the case of massless QCD, it is sufficient to replace $\mu\!\!\!/$ by a commuting (complex) number. For this purpose, we abandon for a moment the case of integer space-time dimensions and follow the conventions from Bern and Morgan in Ref. [62] to define the gamma matrices in $D = 4 - 2\epsilon$ dimensions. In the D-dimensional Dirac algebra,

$$\{\gamma^\mu, \gamma^\nu\} = 2g^{\mu\nu} \tag{2.106}$$

μ, ν are D-dimensional indices and the metric reads $g^{\mu\nu} = \text{diag}(1, -1, -1, -1, -1, \ldots)$. With these conventions, the loop momentum squared reads

$$\ell \cdot \ell = \frac{1}{2}\{\ell, \ell\} = \frac{1}{2}\{\ell_{[4]}, \ell_{[4]}\} + \frac{1}{2}\{\mu\!\!\!/, \mu\!\!\!/\} + \frac{1}{2}\{\ell_{[4]}, \mu\!\!\!/\} + \frac{1}{2}\{\mu\!\!\!/, \ell_{[4]}\}$$
$$= \ell_{[4]} \cdot \ell_{[4]} + \mu \cdot \mu = \ell_{[4]}^2 - \mu^2.$$

consistent with Eq. (2.86). We have exploited the fact that $\mu\!\!\!/$ anti-commutes with any four-dimensional slashed vector. We define γ_5 in $D = 4 - 2\epsilon$ dimensions in the 't Hooft and Veltman scheme [135]:

$$\gamma_5 = i\gamma_0\gamma_1\gamma_2\gamma_3. \tag{2.107}$$

As a consequence $\mu\!\!\!/$ commutes with γ_5:

$$[\mu\!\!\!/, \gamma_5] = 0. \tag{2.108}$$

In order to analyse the action of $\not{\mu}$, we consider an arbitrary Feynman diagram in which a single fermion line contributes with k loop propagators. This can be either a closed quark loop or an external quark that enters the loop. In the latter case, the same arguments that we present now hold also for an arbitrary number of fermion lines. The k propagators will lead to a string of Dirac matrices according to

$$S = (\ell_{1;[4]} + \not{\mu})V_1(\ell_{2;[4]} + \not{\mu})V_2 \ldots V_{k-1}(\ell_{k;[4]} + \not{\mu}). \tag{2.109}$$

V_i are contractions of gamma matrices with external four-dimensional vector currents V_i^μ. Expanding the product, anti-commuting all $\not{\mu}$ to the left through the gamma matrices contracted with four-dimensional vectors and using $\not{\mu}\not{\mu} = -\mu^2$, S exhibits the following structure

$$S = (\gamma)^{2k-1} + \not{\mu}(\gamma)^{2k-2} + \mu^2(\gamma)^{2k-3} + \not{\mu}\mu^2(\gamma)^{2k-4} + \ldots + (\not{\mu})^{k-2\lfloor k/2 \rfloor}(\mu^2)^{\lfloor k/2 \rfloor}(\gamma)^{k-1}.$$

$(\gamma)^m$ collects all terms with m four-dimensional gamma matrices. The relative order of the gamma matrices contracted with four-dimensional vectors remains unchanged, of course. $\lfloor x \rfloor$ is the largest integer less than or equal to x. The notation indicates that $\not{\mu}$ appears only to the power of one for arbitrary number of propagators k. The crucial point is that after anti-commuting the $\not{\mu}$ and multiplying them pairwise always compensates possible sign changes and disguises the non-commuting nature of $\not{\mu}$. This reflects the fact that any pair of two $\not{\mu}$ in Eq. (2.109) is always separated by an odd number of four-dimensional gamma matrices. Hence, if we manage to show that the remaining terms proportional to $\not{\mu}$ vanish without explicitly making use of the non-commuting nature of $\not{\mu}$, then we can replace $\not{\mu}$ by a commuting (complex) number and no explicit representation is required. We will show this separately for the closed quark loop and the mixed quark–gluon loop.

We first note that any remaining $\not{\mu}$ in S is always multiplied with an even number of gamma matrices, while all terms without $\not{\mu}$ involve an odd number of gamma matrices. Hence, we can write schematically

$$S = S^{\text{odd}} + \not{\mu} S^{\text{even}}.$$

To investigate the structure of the closed quark loop, we multiply S with one additional V_k and take the trace of the object. The trace involving $\not{\mu}$ has now an odd number of four-dimensional gamma matrices. Since $\not{\mu}$ freely commutes with γ_5, we can insert $\gamma_5^2 = \mathbb{1}$ in the trace, anti-commute one γ_5 through all four-dimensional gamma matrices and use the cyclic invariance of the trace to show that this contribution vanishes. Since in this argument we did not make use of the non-commuting nature of $\not{\mu}$, we conclude that we can safely compute contributions with a massless closed quark loop replacing $\not{\mu}$ with a commuting (complex) number.

Following the fermion line in the mixed quark–gluon case in both directions out of the loop, S is in general multiplied from both sides with an even number of gamma matrices contracted with purely external momenta and vector currents (remember that the quark–scalar vertex does not involve a Dirac matrix):

$$S^{\text{mixed}} = (\gamma)^{2m_1}(S^{\text{odd}} + \not\mu \, S^{\text{even}})(\gamma)^{2m_2}$$

where m_1 and m_2 are the numbers of external propagators. Since $\not\mu$ commutes with γ_5, it commutes also with the helicity projection operator $\Pi_\pm = \frac{1}{2}(1 \pm \gamma_5)$, i.e. $\Pi_\pm \not\mu = \not\mu \Pi_\pm$. This is not true for the four-dimensional gamma matrices where $\Pi_\pm(\gamma)^{2m} = (\gamma)^{2m}\Pi_\pm$ and $\Pi_\pm(\gamma)^{2m+1} = (\gamma)^{2m+1}\Pi_\mp$ for $m \in \mathbb{N}$ holds. Massless quarks are in helicity eigenstates and in particular, the helicity is conserved along the fermion line. From the explicit form of the external wave functions given in appendix A in Eqs. (A.15) and (A.16) follows that $\Pi_\pm v_\pm = \bar{u}_\pm \Pi_\pm = 0$, $\Pi_\pm v_\mp = v_\mp$ and $\bar{u}_\mp \Pi_\pm = \bar{u}_\mp$. With the convention that all momenta are taken outgoing, the only non-vanishing helicity configurations of an external quark–anti-quark pair is therefore $\bar{u}_\pm S^{\text{mixed}} v_\mp$ (helicity conserving configurations), while $\bar{u}_\pm S^{\text{mixed}} v_\pm$ vanishes (helicity violation along the fermion line). For helicity conserving configurations, all terms proportional to $\not\mu$ automatically vanish since $\Pi_\pm$ commutes with the in total even number of gamma matrices:

$$\bar{u}_\pm(\gamma)^{2m_1}\not\mu \, S^{\text{even}}(\gamma)^{2m_2}v_\mp = \bar{u}_\pm\Pi_\mp(\gamma)^{2m_1}\not\mu \, S^{\text{even}}(\gamma)^{2m_2}v_\mp$$
$$= \bar{u}_\pm(\gamma)^{2m_1}\not\mu \, S^{\text{even}}(\gamma)^{2m_2}\Pi_\mp v_\mp = 0.$$

Again, we did not make any use of the non-commuting properties of $\not\mu$ and we can therefore safely replace it with a commuting (complex) number. From the above derivation follows that the integrand of helicity violating contributions along the fermion line does not vanish, as long as one thinks of $\not\mu$ in terms of a complex number. We checked indeed numerically that with the described approach, the helicity violating contributions are in general non-vanishing, as expected. For our application we may still think of $\not\mu$ as a commuting number because we simply set these helicity violating contributions equal to zero by hand. It is very instructive to convince oneself that in $D_s = 6$ dimensions, there is an explicit representation of $\not\mu$ such that all odd powers of $\not\mu$ cancel independently of the helicity configuration (c.f. Ref. [71] for representations of gamma matrices in higher integer spin dimensions).

A very interesting option to compute also the helicity violating contributions explicitly is to parametrise the integrand in terms of both odd and even powers in μ. In this case, the odd powers are promoted to spurious terms which are artificially present only at the integrand level due to our incorrect interpretation of $\not\mu$ as a commuting (complex) number. The pentagon residue is then not constant any more but has the functional form

$$\bar{e}(\ell) = e^{(0)} + e^{(1)}\mu \tag{2.110}$$

which can be derived in an analogous way as presented in section 2.3, though with the additional (wrong) constraint that the contraction of the epsilon dependent part of the loop momentum with external quantities does not vanish ($u_j \cdot n_\epsilon \neq 0$). In the same spirit, also the boxes, triangles and bubbles receive additional terms proportional to μ

or μ^3 which enter as subtraction terms in lower topologies:

$$\bar{d}(\ell) = \bar{d}_{\text{even}}(\ell) + d^{(5)}\mu + d^{(6)}\alpha_1\mu + d^{(7)}\mu^3 + d^{(8)}\alpha_1\mu^3, \tag{2.111}$$

$$\bar{c}(\ell) = \bar{c}_{\text{even}}(\ell) + c^{(10)}\mu + c^{(11)}\alpha_1\mu + c^{(12)}\alpha_2\mu + c^{(13)}\mu^3, \tag{2.112}$$

$$\bar{b}(\ell) = \bar{b}_{\text{even}}(\ell) + b^{(10)}\mu + b^{(11)}\alpha_1\mu + b^{(12)}\alpha_2\mu + b^{(13)}\alpha_3\mu. \tag{2.113}$$

The parametrisations of $\bar{d}_{\text{even}}(\ell)$, $\bar{c}_{\text{even}}(\ell)$ and $\bar{b}_{\text{even}}(\ell)$ are given in Eqs. (2.70), (2.71) and (2.72). We could numerically confirm that with such an extension, the helicity violating contributions vanish. Although this is computationally highly inefficient — apart from the fact that one computes a result which one knows already in advance, the odd powers need additional integrand evaluations to be disentangled — it is of conceptual interest for a better understanding of integrand reduction techniques.

We conclude with a remark on why this approach is a priori restricted to massless fermions. In the presence of an additional physical mass m, any pair of $\slashed{\mu}$ in Eq. (2.109) would be separated by odd and even numbers of four-dimensional gamma matrices. Anticommuting all $\slashed{\mu}$ in the first position would thus generate contributions with different sign, a fact which cannot be incorporated by a commuting (complex) number. As an example, consider the non-trivial case of three subsequent loop propagators

$$S(m) = (\slashed{\ell}_{1;[4]} + \slashed{\mu} + m)V_1(\slashed{\ell}_{2;[4]} + \slashed{\mu} + m)V_2(\slashed{\ell}_{1;[4]} + \slashed{\mu} + m).$$

Collecting only the terms with two powers in $\slashed{\mu}$ and one power in m, and doing the same algebra as in the massless case results in a contribution $\mu^2 m V_1 V_2$ while interpreting $\slashed{\mu}$ as a commuting number from the beginning on would give $3\mu^2 m V_1 V_2$. Thus the generated integrand is different. In order to include also massive fermions, an explicit anti-commuting representation of $\slashed{\mu}$ is required. A suggestion to include also massive external quarks in this formalism is the following: One replaces $\slashed{\mu}$ by $i\mu\gamma_5$ and computes the coefficients $d^{(4)}$, $c^{(9)}$ and $b^{(9)}$ by disentangling both odd *and* even powers in μ, as described above. An internal "true" massive fermion propagator has then a numerator of the form $\slashed{\ell}_{[4]} + i\mu\gamma_5 + m$. The completeness relations that reconstructs such a numerator require an extension of the conventional massive four-spinors that solve the four- dimensional Dirac equation. A specific representation for such modified four-spinors is given in Eqs. (A.25) – (A.28) in appendix A. The inclusion of wave function renormalisation and mass renormalisation proceeds along the lines described in Ref. [71].

Outline of the one-loop computation: The computation of one-loop primitive amplitudes in massless QCD is now a two step procedure. First one computes the cut-constructible part in strictly four dimensions setting $\mu^2 = 0$. Note that there is no pentagon contribution, as argued above. The actual computation is explained for any massless amplitude in great detail in section 2.5. In a second step, the rational part is computed independently, now including pentagons but still in four dimensions, however, with a massive loop momentum. The particle content inside the loop is now different: gluons are replaced with massive scalars, and massless quarks with massive quarks. This difference is manifest at the level of the tree-level input with which one reconstructs the integrand. The inclusion of pentagons and the different form of the integrand lead to an algorithm which is different from the cut-constructible part. How this works in detail is explained in section 2.6. Since the external kinematics is the same for both cut-constructible and rational part, many quantities need to be computed only once, like for example the transverse vectors n_i^μ or off-shell currents that involve only external partons. More details about the latter are described in appendix B.1.

2.5. Computation of the cut-constructible part

As explained in the previous section, we compute the cut-constructible part without pentagon contributions. The numerator basis given in Eqs. (2.78) – (2.81) simplifies therefore to

$$\bar{d}^{[4]}_{i_1 i_2 i_3 i_4}(\ell) = \text{Res}_{[i_1 i_2 i_3 i_4]}\Big(\mathcal{G}_n(\ell)\Big), \tag{2.114}$$

$$\bar{c}^{[4]}_{i_1 i_2 i_3}(\ell) = \text{Res}_{[i_1 i_2 i_3]}\left(\mathcal{G}_n(\ell) - \sum_{[i_1|i_4]} \frac{\bar{d}^{[4]}_{i_1 i_2 i_3}(\ell)}{d_{i_1} d_{i_2} d_{i_3} d_{i_4}}\right), \tag{2.115}$$

$$\bar{b}^{[4]}_{i_1 i_2}(\ell) = \text{Res}_{[i_1 i_2]}\left(\mathcal{G}_n(\ell) - \sum_{[i_1|i_4]} \frac{\bar{d}^{[4]}_{i_1 i_2 i_3}(\ell)}{d_{i_1} d_{i_2} d_{i_3} d_{i_4}} - \sum_{[i_1|i_3]} \frac{\bar{c}^{[4]}_{i_1 i_2 i_3}(\ell)}{d_{i_1} d_{i_2} d_{i_3}}\right). \tag{2.116}$$

The four-dimensional tensor structure of $\bar{d}^{[4]}_{i_1 i_2 i_3 i_4}$, $\bar{c}^{[4]}_{i_1 i_2 i_3}$ and $\bar{b}^{[4]}_{i_1 i_2}$ is recovered from Eqs. (2.70), (2.71) and (2.72) setting $\mu^2 = 0$ which coincides with the one given in Ref. [72]. The superscript [4] indicates that the functions involve only the four-dimensional tensor structures. The actual computation of the integral coefficients proceeds similar as in Ref. [72]. We describe the main features of the algorithm and emphasise how we solve the occurring systems of equations by parametrising the integrand in terms of orthogonal functions.

2.5.1. The box coefficients

In order to compute the box residue $\bar{d}^{[4]}_{i_1 i_2 i_3 i_4}$ four propagators need to be set on-shell. This generates four equations according to the unitarity constraints $d_{i_1} = d_{i_2} = d_{i_3} = d_{i_4} = 0$. It is straightforward to apply these constraints in the master equation for the loop momentum decomposition (2.54) simply skipping those terms that involve inverse propagators. We arrive at

$$\ell^\mu = V_4^\mu + (\ell \cdot n_1)\, n_1^\mu = V_4^\mu + \alpha_1\, n_1^\mu \tag{2.117}$$

with

$$V_4^\mu = -\frac{1}{2} q_1^2 v_1^\mu + \frac{1}{2} q_3^2 v_2^\mu + \frac{1}{2} \left(q_4^2 - q_3^2 \right) v_3^\mu. \tag{2.118}$$

Remember that all masses are set equal to zero and that in our parametrisation $q_2 = 0$, c.f. Eq. (2.5). The four unitarity constraints lead to four equations of the form

$$d_{i_r}(\ell) = (\ell - q_r)^2 = (V_4 + \alpha_1 n_1 - q_r)^2 = V_4^2 + \alpha_1^2 - 2V_4 \cdot q_r + q_r^2 \overset{!}{=} 0. \tag{2.119}$$

Since $V_4 \cdot q_r = \frac{1}{2} q_r^2$, c.f. Eq (2.56), the four unitarity constraints are equivalent to a single quadratic equation in α_1

$$V_4^2 + \alpha_1^2 = 0 \tag{2.120}$$

which is independent of r and with two solutions

$$\alpha_1 = \pm \sqrt{-V_4^2} \equiv \pm \kappa. \tag{2.121}$$

Hence, setting four propagators on the mass shell fixes the value of α_1 and leads to two (complex) on-shell loop momenta

$$\ell_\pm^\mu = V_4^\mu \pm \kappa\, n_1^\mu \tag{2.122}$$

for which by construction $(\ell_\pm - q_r)^2 = 0$ holds. For $\mu^2 = 0$ in Eq. (2.70), there are only two independent tensor structures in four dimensions

$$\bar{d}_{i_1 i_2 i_3 i_4}^{[4]}(\ell) = d^{(0)} + d^{(1)}\alpha_1. \tag{2.123}$$

For this reason, the two solutions $\ell_\pm$ are sufficient to determine the coefficients $d^{(0)}$ and $d^{(1)}$: Evaluating the numerator function $\bar{d}_{i_1 i_2 i_3 i_4}^{[4]}(\ell)$ twice, once for $\ell = \ell_+$ and once for $\ell = \ell_-$, generates a system of two equations whose solution can immediately be written down as

$$d^{(0)} = \frac{1}{2} \left(\bar{d}_{i_1 i_2 i_3 i_4}^{[4]}(\ell_+) + \bar{d}_{i_1 i_2 i_3 i_4}^{[4]}(\ell_-) \right), \tag{2.124}$$

$$d^{(1)} = \frac{1}{2\,\kappa} \left(\bar{d}_{i_1 i_2 i_3 i_4}^{[4]}(\ell_+) - \bar{d}_{i_1 i_2 i_3 i_4}^{[4]}(\ell_-) \right). \tag{2.125}$$

As discussed in section 2.3 in detail, $d^{(0)}$ is the desired integral coefficient of the topology $C_R = \{i_1, i_2, i_3, i_4\}$. Both $d^{(0)}$, $d^{(1)}$ and also the transverse vector n_1^μ of this distinct topology will be needed later on in the triangle and bubble case for the correct reconstruction of the box integrand to subtract contributions with common propagators. In the four-dimensional box case no subtractions are necessary and $\bar{d}_{i_1 i_2 i_3 i_4}^{[4]}(\ell)$ is directly the residue $\mathrm{Res}_{[i_1 i_2 i_3 i_4]}(\mathcal{G}_n(\ell))$, c.f. Eq. (2.114). We stress once again that $\bar{d}_{i_1 i_2 i_3 i_4}^{[4]}(\ell_\pm)$ is computed by taking products of four tree-level amplitudes according to Eq. (2.77).

2.5.2. The triangle coefficients

In order to compute the triangle residue $\bar{c}^{[4]}_{i_1 i_2 i_3}(\ell)$ three propagators need to be set on-shell. This generates three equations according to the unitarity constraints $d_{i_1} = d_{i_2} = d_{i_3} = 0$. Plugging these constraints into Eq. (2.54), the triangle on-shell loop momentum reads

$$\ell^\mu = V_3^\mu + (\ell \cdot n_1)\, n_1^\mu + (\ell \cdot n_2)\, n_2^\mu = V_3^\mu + \alpha_1\, n_1^\mu + \alpha_2\, n_2^\mu \tag{2.126}$$

with

$$V_3^\mu = -\frac{1}{2}\, q_1^2\, v_1^\mu + \frac{1}{2}\, q_3^2\, v_2^\mu\,. \tag{2.127}$$

The three unitarity constraints lead to three equations of the form

$$d_{i_r}(\ell) = (\ell - q_r)^2 = (V_3 + \alpha_1 n_1 + \alpha_2 n_2 - q_r)^2$$
$$= V_3^2 + \alpha_1^2 + \alpha_2^2 - 2V_3 \cdot q_r + q_r^2 \overset{!}{=} 0. \tag{2.128}$$

Since $V_3 \cdot q_r = \frac{1}{2}q_r^2$, c.f. Eq (2.56), the three unitarity constraints are equivalent to a single quadratic form in α_1 and α_2

$$\alpha_1^2 + \alpha_2^2 = -V_3^2 \tag{2.129}$$

which is, like in the box case, independent of r. The crucial difference to the box case, however, is that there are now infinitely many solutions that fulfil the quadratic form (2.129). This is expected because there are only three equations that constrain four components of the loop momentum, hence, the on-shell loop momentum must still depend on one free parameter. We can now exploit this freedom and choose as many different pairs of (α_1, α_2) fulfilling Eq. (2.129) as there are independent tensor structures in the triangle numerator. Evaluating $\bar{c}^{[4]}_{i_1 i_2 i_3}(\ell)$ with the corresponding loop momenta then generates a system of equations whose solution contains precisely the triangle integral coefficient and the coefficients of the spurious terms. Remember that we compute the triangle residue by taking a product of three tree-level amplitudes and then we subtract those box integrands with common propagators. With the integral coefficients $d^{(0)}$, $d^{(1)}$ and the transverse box vector n_1^μ from the preceding box evaluation, the subtraction term has the form

$$\sum_{i \notin C_3} \frac{d_i^{(0)} + d_i^{(1)}(\ell_\Delta - q_i) \cdot n_{1,i}}{(\ell_\Delta - q_i)^2}$$

where the triangle subscript at ℓ_Δ shall emphasise that this is the on-shell triangle momentum, i is the index of the common propagators and q_i is the corresponding momentum parametrisation. Setting $\mu^2 = 0$ in Eq. (2.71), one recovers the seven independent tensor structures of the four-dimensional case

$$\bar{c}^{[4]}_{i_1 i_2 i_3}(\ell) = c^{(0)} + c^{(1)} \alpha_1 + c^{(2)} \alpha_2 + c^{(3)} (\alpha_1^2 - \alpha_2^2) + c^{(4)} \alpha_1 \alpha_2$$
$$+ c^{(5)} \alpha_1^2 \alpha_2 + c^{(6)} \alpha_1 \alpha_2^2 \,. \tag{2.130}$$

In Ref. [72] it is shown, that the corresponding system of equations leads to Vander-monde type matrices which are known to be unstable to invert. One technique for a more stable inversion is *discrete Fourier transformation (DFT)*. This technique exploits the polynomial structure of the integrand as input to disentangle the coefficients $c^{(i)}$ in Eq. (2.130) from each other. The idea is to express the monomials $\alpha_1^p \alpha_2^q$ via orthogonal basis functions. The projection onto these functions, then, allows one directly to read off the coefficients $c^{(i)}$. The fundamentals of this technique are explained in standard textbooks on numerics like for example [136]. First applications in the context of inte-grand reduction can be found in Refs. [84, 137]. Since some of the basis functions and the parametrisation of the integrand differ from [84, 137], we will derive all formulae explicitly.

Let us assume that we can parametrise the integrand in terms of a finite set of $(p+1)$ orthogonal functions $f_n(\ell_j)$ with $n = 0, \ldots, p$ for which a completeness relation holds

$$\sum_{j=0}^{p} f_m(\ell_j) f_n(\ell_j) = \delta_{mn} \,. \tag{2.131}$$

ℓ_j is a suggestive notation to indicate that the loop momentum ℓ depends on a discrete index j. The integrand can then be expanded in terms of these orthogonal functions according to

$$\bar{c}(\ell_j) = \sum_{m=0}^{p} \tilde{c}^{(m)} f_m(\ell_j). \tag{2.132}$$

The tilde at $\tilde{c}^{(m)}$ indicates that the coefficients need not automatically be identified with the $c^{(m)}$ in Eq. (2.130). Depending on the parametrisation one could also think of distinct linear combinations $\tilde{c}^{(n)} = \sum_m a_{nm} c^{(m)}$. Due to the orthogonality we can directly project out a distinct coefficient $\tilde{c}^{(n)}$ by means of

$$\sum_{i=0}^{p} \bar{c}(\ell_i) f_n(\ell_i) = \sum_{i=0}^{p} \sum_{m=0}^{p} \tilde{c}^{(m)} f_m(\ell_i) f_n(\ell_i) = \sum_{m=0}^{p} \tilde{c}^{(m)} \delta_{mn} = \tilde{c}^{(n)}. \tag{2.133}$$

It turns out that we can indeed find an orthogonal set of basis functions with which we can parametrise the integrand.

Let us come back to the unitarity constraint $\alpha_1^2 + \alpha_2^2 = -V_3^2$ from Eq. (2.129) which a pair (α_1, α_2) must fulfil. The first important observation is that V_3^2 factorises for arbitrary outflow momenta $\{k_1, k_2, k_3\}$ into

$$V_3^2 = \frac{k_1^2\, k_2^2\, k_3^2}{4\,\Delta} \tag{2.134}$$

where $\Delta = \Delta(k_1, k_2)$ is the two-particle Gram determinant defined in Eq. (2.25). It follows that if at least one of the external outflow momenta is light-like, then $V_3^2 = 0$ and the unitarity condition shrinks to

$$\alpha_1^2 + \alpha_2^2 = 0\,. \tag{2.135}$$

A non-trivial solution of this equation is necessarily complex and can be parametrised as $(\alpha_1, \alpha_2) = (z, \pm iz)$ with $z \in \mathbb{C}$. In terms of the discrete Fourier transformation, the set of basis functions can be chosen real for the case $V_3^2 \neq 0$ while it must be chosen complex for the massless case. Since the parametrisation of the integrand is different, we will treat both cases separately.

For $V_3^2 \neq 0$, the unitarity constraint $\alpha_1^2 + \alpha_2^2 = -V_3^2$ describes a circle of radius

$$\kappa = \sqrt{-V_3^2} \tag{2.136}$$

which can be parametrised as

$$\alpha_1 = \kappa \sin(x) \tag{2.137}$$
$$\alpha_2 = \kappa \cos(x) \tag{2.138}$$

and $x \in \mathbb{R}$. As real, orthogonal basis functions we choose the trigonometric functions

$$f_m(j) = \left\{ \sin\left(\frac{2\pi}{2p+1} j\, m\right),\ \cos\left(\frac{2\pi}{2p+1} j\, m\right) \right\} \tag{2.139}$$

with $p \in \mathbb{N}$ and $j, m = \{-p, -p+1, \ldots, p-1, p\}$. They obey the completeness relation

$$\sum_{j=-p}^{p} \sin\left(\frac{2\pi}{2p+1} j\, m\right) \sin\left(\frac{2\pi}{2p+1} j\, n\right) = \begin{cases} \frac{2p+1}{2}\delta_{mn} & \text{for } \{m,n\} \neq 0 \\ 0 & \text{for } \{m,n\} = 0 \end{cases}, \tag{2.140}$$

$$\sum_{j=-p}^{p} \cos\left(\frac{2\pi}{2p+1} j\, m\right) \cos\left(\frac{2\pi}{2p+1} j\, n\right) = \begin{cases} \frac{2p+1}{2}\delta_{mn} & \text{for } \{m,n\} \neq 0 \\ 2p+1 & \text{for } \{m,n\} = 0 \end{cases}, \tag{2.141}$$

$$\sum_{j=-p}^{p} \sin\left(\frac{2\pi}{2p+1} j\, m\right) \cos\left(\frac{2\pi}{2p+1} j\, n\right) = 0\,. \tag{2.142}$$

This choice of functions fits perfectly with the unitarity constraints $\alpha_1 = \kappa \sin(x)$ and $\alpha_2 = \kappa \cos(x)$ in Eqs. (2.137) and (2.138) setting

$$x = \frac{2\pi}{2p+1}\, j. \tag{2.143}$$

It implies that we choose $(2p + 1)$ points for (α_1, α_2) equally distributed on a circle of radius κ. The loop momentum can therefore be written as

$$\ell_j^\mu = V_3^\mu + \kappa \, \sin\left(\frac{2\pi}{2p+1}j\right) n_1^\mu + \kappa \, \cos\left(\frac{2\pi}{2p+1}j\right) n_2^\mu \qquad (2.144)$$

which makes the dependence on the discrete parameter j manifest. Using trigonometric identities, we rewrite the monomials of $\alpha_1^p \alpha_2^q$ where $0 \leq p + q \leq 3$ in the integrand $\bar{c}(\ell_j)$ in Eq. (2.130) by means of orthogonal basis functions. In other words, we express $\sin^p(x)\cos^q(x)$ in terms of $\sin(mx)$ and $\cos(nx)$ where $m, n \leq 3$. The integrand thus reads:

$$\begin{aligned}
\bar{c}(\ell_x) = {} & c^{(0)} + [\kappa c^{(1)} + \frac{\kappa^3}{4}c^{(6)}]\sin(x) + [\kappa c^{(2)} + \frac{\kappa^3}{4}c^{(5)}]\cos(x) \\
& + \frac{\kappa^2}{2}c^{(4)}\,\sin(2x) - \kappa^2 c^{(3)}\,\cos(2x) \\
& + \frac{\kappa^3}{4}c^{(6)}\,\sin(3x) - \frac{\kappa^3}{4}c^{(5)}\,\cos(3x)
\end{aligned} \qquad (2.145)$$

which is precisely the form of Eq. (2.132). The projection onto the integral coefficients is now straightforward. We find

$$P[0] = \frac{1}{2p+1}\sum_{j=-p}^{p} \bar{c}(\ell_j) = c^{(0)}$$

$$P[1] = \frac{2}{2p+1}\sum_{j=-p}^{p} \sin\left(\frac{2\pi}{2p+1}j\right)\bar{c}(\ell_j) = \kappa\,c^{(1)} + \frac{\kappa^3}{4}c^{(6)}$$

$$P[2] = \frac{2}{2p+1}\sum_{j=-p}^{p} \cos\left(\frac{2\pi}{2p+1}j\right)\bar{c}(\ell_j) = \kappa\,c^{(2)} + \frac{\kappa^3}{4}c^{(5)}$$

$$P[3] = \frac{2}{2p+1}\sum_{j=-p}^{p} \sin\left(\frac{2\pi}{2p+1}2\,j\right)\bar{c}(\ell_j) = -\kappa^2 c^{(3)}$$

$$P[4] = \frac{2}{2p+1}\sum_{j=-p}^{p} \cos\left(\frac{2\pi}{2p+1}2\,j\right)\bar{c}(\ell_j) = \frac{\kappa^2}{2}c^{(4)}$$

$$P[5] = \frac{2}{2p+1}\sum_{j=-p}^{p} \sin\left(\frac{2\pi}{2p+1}3\,j\right)\bar{c}(\ell_j) = -\frac{\kappa^3}{4}c^{(5)}$$

$$P[6] = \frac{2}{2p+1}\sum_{j=-p}^{p} \cos\left(\frac{2\pi}{2p+1}3\,j\right)\bar{c}(\ell_j) = \frac{\kappa^3}{4}c^{(6)} \qquad (2.146)$$

Up to numerical prefactors, we get immediately $c^{(0)}, c^{(3)}, c^{(4)}, c^{(5)}, c^{(6)}$. From these projections we can further get $c^{(1)}$ and $c^{(2)}$ via the linear combinations

$$\kappa \, c^{(1)} = P[1] - P[6],$$
$$\kappa \, c^{(2)} = P[2] + P[5].$$

Dividing by the appropriate powers of κ and the numerical factors, we get all coefficients $c^{(i)}$. Since there are seven independent coefficients $c^{(i)}$, we have to evaluate the integrand $\bar{c}(\ell)$ at least seven times. Choosing $p = 3$ leads exactly to the seven required basis functions. Choosing $p \geq 4$, allows a projection on additional basis functions, for example with $p = 4$, these are $\sin(4x)$ and $\cos(4x)$. Such functions necessarily belong to a rank four tensor structure which is not present in the triangle case. Nevertheless, we can sample more than seven points and project on such higher rank coefficients to check numerically that they vanish. We did such investigations and found indeed that all higher rank coefficients vanished. However, a relation between the accuracy of the coefficients $c^{(0)}, \ldots, c^{(6)}$ and the "goodness" of the zero that we have computed could not be establish. One can further argue that the Fourier decomposition with more sampled points leads to better accuracy for the non-vanishing integral coefficients. As we will see in the computation of the rational part where we also applied the DFT, this is indeed true. Yet, for the cut-constructible triangles we do not find any improvement using higher modes. For these reasons, we apply the DFT always with $p = 3$ leading to the minimum of seven integrand evaluations.

For $V_3^2 = 0$, the unitarity constraint $\alpha_1^2 + \alpha_2^2 = 0$ leads to complex solutions whose general form reads

$$\alpha_1 = z = r_0 \exp(ix) \tag{2.147}$$
$$\alpha_2 = \pm iz = i\sigma \, r_0 \exp(ix) \tag{2.148}$$

with an arbitrary complex number $z = r_0 \exp(ix) \,|\, r_0, x \in \mathbb{R}$ and $\sigma = \pm 1$. Note that for any arbitrary z, there are always two independent solutions of the loop momentum labelled by $\sigma = \pm 1$. In contrast to the real case, there are now two free parameters, x and r_0. We will comment below on a suitable choice for the radius r_0. We choose as orthogonal complex basis functions

$$f_m(j) = \exp\left(\frac{2\pi i}{p+1} j \, m\right) \tag{2.149}$$

with $p \in \mathbb{N}$, $j, m = \{0, 1, \ldots, p-1, p\}$ and i the imaginary unit. They obey the completeness relation

$$\sum_{j=0}^{p} \exp\left(\frac{2\pi i}{p+1} j \, m\right) \exp\left(\frac{2\pi i}{p+1} j \, n\right) = (p+1)\delta_{mn}. \tag{2.150}$$

This choice of functions matches automatically the unitarity constraints $\alpha_1 = r_0 \exp(ix)$

and $\alpha_2 = i\sigma\, r_0 \exp(ix)$ in Eqs. (2.147) and (2.148) setting

$$x = \frac{2\pi i}{p+1}\, j. \tag{2.151}$$

The loop momentum can therefore be written as

$$\ell^\mu_{j,\sigma} = V_3^\mu + r_0 \exp\left(\frac{2\pi i}{p+1} j\right) n_1^\mu + i\sigma r_0 \exp\left(\frac{2\pi i}{p+1} j\right) n_2^\mu. \tag{2.152}$$

Note that ℓ depends also on σ which labels two independent solutions of the loop momentum for a given j. It is straightforward to express the integrand in terms of the basis functions. We get

$$\begin{aligned}
\bar{c}\,(\ell_{x,\sigma}) = \ & c^{(0)} + [\ c^{(1)} + \sigma\, i\, c^{(2)}]\, r_0 \exp(ix)\\
& + [2\, c^{(3)} + \sigma\, i\, c^{(4)}]\, r_0^2 \exp(2ix)\\
& + [-c^{(6)} + \sigma\, i\, c^{(5)}]\, r_0^3 \exp(3ix).
\end{aligned}$$

If we were interested in $c^{(0)}$ only, the projection onto the basis functions $\exp(nix)$ with $n = 0 \ldots 3$ would be sufficient. Since we have to construct the complete triangle integrand for the computation of the bubble coefficients, we have to disentangle also the remaining coefficients. This is achieved by a subsequent projection with respect to σ. The full disentanglement of the coefficients finally reads

$$c^{(0)} = \frac{1}{2}\frac{1}{p+1} \sum_{\sigma=\pm 1}\sum_{j=0}^{p} \bar{c}\,(l_{j,\sigma})$$

$$c^{(1)} = \frac{1}{2}\frac{1}{p+1} \sum_{\sigma=\pm 1}\sum_{j=0}^{p} \bar{c}\,(l_{j,\sigma})\, \frac{1}{r_0}\, \exp\left(\frac{2\pi i}{p+1}\, j\right)$$

$$ic^{(2)} = \frac{1}{2}\frac{1}{p+1} \sum_{\sigma=\pm 1}\sum_{j=0}^{p} \bar{c}\,(l_{j,\sigma})\, \frac{\sigma}{r_0}\, \exp\left(\frac{2\pi i}{p+1}\, j\right)$$

$$2c^{(3)} = \frac{1}{2}\frac{1}{p+1} \sum_{\sigma=\pm 1}\sum_{j=0}^{p} \bar{c}\,(l_{j,\sigma})\, \frac{1}{r_0^2}\, \exp\left(\frac{2\pi i}{p+1}\, 2j\right)$$

$$ic^{(4)} = \frac{1}{2}\frac{1}{p+1} \sum_{\sigma=\pm 1}\sum_{j=0}^{p} \bar{c}\,(l_{j,\sigma})\, \frac{\sigma}{r_0^2}\, \exp\left(\frac{2\pi i}{p+1}\, 2j\right)$$

$$ic^{(5)} = \frac{1}{2}\frac{1}{p+1} \sum_{\sigma=\pm 1}\sum_{j=0}^{p} \bar{c}\,(l_{j,\sigma})\, \frac{\sigma}{r_0^3}\, \exp\left(\frac{2\pi i}{p+1}\, 3j\right)$$

$$-c^{(6)} = \frac{1}{2}\frac{1}{p+1} \sum_{\sigma=\pm 1}\sum_{j=0}^{p} \bar{c}\,(l_{j,\sigma})\, \frac{1}{r_0^3}\, \exp\left(\frac{2\pi i}{p+1}\, 3l\right)$$

The minimal choice of p to disentangle the system is $p = 3$ corresponding to four Fourier modes. With the two solutions $\sigma = \pm 1$, we finally have $2(p+1)$ integrand evaluations[3], i.e. the minimal number of integrand evaluations is eight which is one more than the number of independent structures $c^{(i)}$. With the information of the additional integrand evaluation, one gets an eighth coefficient $c^{(\star)}$

$$c^{(\star)} = \sum_{\sigma = \pm 1} \sum_{j=0}^{p} \bar{c} \left(\ell_{j,\sigma} \right) \sigma \,. \tag{2.153}$$

This coefficient necessarily vanishes, because there is not a structure like $c^{(\star)} \sigma$ present in the general form of the integrand in Eq. (2.130). We could numerically verify that this projection is indeed zero. However, we did not use this information to draw conclusions about the numerical accuracy.

We conclude the massless triangle case with a remark on the free parameter r_0. Since the transverse vectors n_i^μ are dimensionless quantities, all $\alpha_i = \ell \cdot n_i$ have the dimension of the loop momentum, i.e. mass dimension one. To avoid large numerical cancellations, we relate $r_0 = |\alpha_i|$ to a scale of the problem. A natural and consistent choice is the centre of mass energy $r_0 = \sqrt{s}$.

We mention at this point that we have to store all integrand coefficients $c^{(i)}$ and the two transverse triangle vectors n_1^μ and n_2^μ for the correct reconstruction of the triangle integrand in the computation of the bubble coefficients.

2.5.3. The bubble coefficients

In order to compute the bubble residue $\bar{b}_{i_1 i_2}^{[4]}(\ell)$ two propagators need to be set on-shell. This generates two equations according to the unitarity constraints $d_{i_1} = d_{i_2} = 0$. Plugging these constraints into Eq. (2.54), the bubble on-shell loop momentum reads

$$\begin{aligned} \ell^\mu &= V_2^\mu + (\ell \cdot n_1) \, n_1^\mu + (\ell \cdot n_2) \, n_2^\mu + (\ell \cdot n_3) \, n_3^\mu \\ &= V_2^\mu + \alpha_1 \, n_1^\mu + \alpha_2 \, n_2^\mu + \alpha_3 \, n_3^\mu \end{aligned} \tag{2.154}$$

with

$$V_2^\mu = -\frac{1}{2} \, q_1^2 \, v_1^\mu \,. \tag{2.155}$$

The two unitarity constraints lead to two equations of the form

$$\begin{aligned} d_{i_r}(\ell) &= (\ell - q_r)^2 = (V_2 + \alpha_1 n_1 + \alpha_2 n_2 + \alpha_3 n_3 - q_r)^2 \\ &= V_2^2 + \alpha_1^2 + \alpha_2^2 + \alpha_3^2 - 2 V_2 \cdot q_r + q_r^2 \overset{!}{=} 0 \end{aligned} \tag{2.156}$$

---

[3]One can consider $\sigma = \pm 1$ also as a second Fourier projection with $p = 1$ leading to two Fourier modes. Such double Fourier projections are explained in detail in section 2.5.3.

Since $V_2 \cdot q_r = \frac{1}{2} q_r^2$, c.f. Eq. (2.56), the two unitarity constraints are equivalent to a single quadratic form in α_1, α_2 and α_3

$$\alpha_1^2 + \alpha_2^2 + \alpha_3^2 = -V_2^2. \tag{2.157}$$

Note that $V_2^2 = k_1^2/4 \neq 0$ because we discard all on-shell bubbles with massless external legs. The triple of numbers $\vec{\alpha} = (\alpha_1, \alpha_2, \alpha_3)$ is restricted to a sphere with radius

$$\kappa = \sqrt{-V_2^2} \tag{2.158}$$

which can be parametrised with spherical coordinates:

$$\alpha_1 = \kappa \sin(x) \cos(y) \tag{2.159}$$
$$\alpha_2 = \kappa \sin(x) \sin(y) \tag{2.160}$$
$$\alpha_3 = \kappa \cos(x) \tag{2.161}$$

As in the previous section we recover the four-dimensional integrand structure by setting $\mu^2 = 0$ in Eq. (2.72)

$$\bar{b}_{i_1 i_2}^{[4]}(\ell) = b^{(0)} + b^{(1)} \alpha_1 + b^{(2)} \alpha_2 + b^{(3)} \alpha_3 + b^{(4)} \left(\alpha_1^2 - \alpha_3^2\right) + b^{(5)} \left(\alpha_2^2 - \alpha_3^2\right)$$
$$+ b^{(6)} \alpha_1 \alpha_2 + b^{(7)} \alpha_1 \alpha_3 + b^{(8)} \alpha_2 \alpha_3. \tag{2.162}$$

Since there are no tadpoles present in our computation which would require a subtraction of the bubble integrand, it is sufficient to disentangle the integral coefficient $b^{(0)}$ from the rest. All other coefficients of the spurious terms $b^{(1)}, \ldots, b^{(8)}$ are not explicitly needed. We will discuss first, how one can compute $b^{(0)}$ with DFT methods as it is implemented in NJET . Subsequently we comment on possible improvements.

In order to apply a DFT, we use the same basis of orthogonal functions as in the triangle case with $V_3^2 \neq 0$ listed in Eq. (2.139). The main difference is that we have to perform two independent Fourier projections. In terms of the parametrisation in Eqs. (2.159) – (2.161) this is one projection with respect to the functions in x and the other one with respect to the functions in y. The independence of both x and y implies that we have to take the direct product of the two function bases. For a total of $(2p_x + 1)$ basis functions f_x and $(2p_y + 1)$ basis functions f_y with $p_x, p_y \in \mathbb{N}$ this is a total of $(2p_x + 1)(2p_y + 1)$ combinations which can be listed as follows

$$f_x \otimes f_y \equiv \{1, \sin(x), \cos(x), \sin(2x), \cos(2x), \ldots, \sin(p_x x), \cos(p_x x),$$
$$\sin(y), \sin(x)\sin(y), \cos(x)\sin(y), \ldots, \cos(p_x x)\cos(p_y y)\} \tag{2.163}$$

with

$$x = \frac{2\pi}{2p_x + 1} j_x \quad \text{and} \quad y = \frac{2\pi}{2p_y + 1} j_y \tag{2.164}$$

where $j_x \in \{-p_x, \ldots, p_x\}$ and $j_y \in \{-p_y, \ldots, p_y\}$. The explicit form of the loop momentum therefore reads

$$
\ell_{j_x,j_y} = V_2^{\mu} + \kappa \sin\left(\frac{2\pi}{2p_x + 1} j_x\right) \cos\left(\frac{2\pi}{2p_y + 1} j_y\right) n_1^{\mu}
$$

$$
+ \kappa \sin\left(\frac{2\pi}{2p_x + 1} j_x\right) \sin\left(\frac{2\pi}{2p_y + 1} j_y\right) n_2^{\mu} + \kappa \cos\left(\frac{2\pi}{2p_x + 1} j_x\right) n_3^{\mu}. \quad (2.165)
$$

Applying again trigonometric identities we can express the integrand by means of basis functions from $f_x \otimes f_y$

$$
\bar{b}_{i_1 i_2}^{[4]}(\ell_{x,y}) = \quad [b^{(0)} - \frac{\kappa^2}{4}(b^{(7)} + b^{(8)})]
$$

$$
+ [\kappa\, b^{(1)}] \sin(x) \cos(y) + [\kappa b^{(2)}] \sin(x) \sin(y) + [\kappa b^{(3)}] \cos(x)
$$

$$
+ [-\frac{\kappa^2}{4} b^{(4)}] \sin(2y) + [\frac{\kappa^2}{4}(b^{(7)} - b^{(8)})] \cos(2y)
$$

$$
+ [\frac{\kappa^2}{2} b^{(5)}] \sin(2x) \cos(y) + [\frac{\kappa^2}{2} b^{(6)}] \sin(2x) \sin(y)
$$

$$
+ [-\frac{3\kappa^2}{4}(b^{(7)} + b^{(8)})] \cos(2x)
$$

$$
+ [\frac{\kappa^2}{4} b^{(4)} + \frac{\kappa^2}{4}(-b^{(7)} + b^{(8)})] \cos(2x) \cos(2y). \quad (2.166)
$$

Although the integrand is now written in terms of ten basis functions $\in f_x \otimes f_y$, for the disentanglement of all coefficients, only a projection onto nine basis functions is necessary. To disentangle $b^{(0)}$ exclusively, it is sufficient to project out the coefficients of the functions ~ 1 and $\sim \cos(2x)$, and to take an appropriate linear combination thereof. The final answer reads

$$
b^{(0)} = \frac{1}{(2p_x + 1)(2p_y + 1)} \sum_{j_x=-p_x}^{p_x} \sum_{j_y=-p_y}^{p_y} \bar{b}_{i_1 i_2}^{[4]}(\ell_{j_x,j_y}) \left(1 - \frac{2}{3} \cos\left(\frac{2\pi}{2p_x + 1} 2j_x\right)\right). \quad (2.167)
$$

The minimal choice of Fourier modes within this basis to get $b^{(0)}$ exclusively is $p_x = 2$ and $p_y = 1$. This leads to a total of 15 necessary evaluations of the integrand $\bar{b}(\ell_{j_x,j_y})$. Since there are only nine independent parameters the system of equations is in this case overconstrained. Yet, at the time when the NGLUON library was developed, we found that this method led to excellent numerical stability. We give in Appendix D.1 an alternative method with which the integral coefficient $b^{(0)}$ can be computed with only four integrand evaluations. This is an improvement of almost a factor of four. We have, however, not tested this alternative on numerical accuracy. In case the accuracy is worse than in the present implementation, we have to switch to quadruple precision for a larger number of phase space points. The overall runtime performance can then even be worse. A careful analysis of this issue is postponed for a future release of NJET .

We conclude this chapter with some comments on why the DFT fits so well in the triangle case and why it is quite cumbersome in the bubble case. The circle that we used to constrain α_1 and α_2 in the triangles is continuously parametrised by the one-parameter $U(1)$ Lie group. A special property of $U(1)$ is that for every $n \in \mathbb{N}$, the cyclic group Z_n forms a discrete subgroup of $U(1)$. Z_n can be used to describe n equidistantly distributed points on a circle which is precisely the underlying symmetry group of the DFT. The bottom line is that we can always adapt n to the highest power of the polynomial in question as long as the polynomial appears effectively only within one parameter (remember that α_1 and α_2 are not independent). For the bubble case, the natural generalisation would suggest to choose n equidistantly distributed points on a three dimensional sphere, however, then, n cannot be chosen arbitrarily any more, it is restricted to $n = 4, 6, 8, 12, 20$ — the vertices of the platonic solids described by the tetraeder, hexaeder and ikosaeder group, discrete subgroups of $SO(3)$. As we show in Appendix D.1 they lead to efficient evaluation of the integral coefficient $b^{(0)}$ but it is not general enough to project out the spurious terms (in case they are needed for tadpoles). Another candidate might seem spherical harmonics Y_{lm} as basis functions. For $l = 0, 1, 2$ and $m = -l, \ldots, +l$, we would indeed have nine basis functions which is exactly the number of independent coefficients $b^{(i)}$. The problem, however, is that there is no discrete version of spherical harmonics. As a consequence, the completeness relation involves always an infinite set of functions and a projection onto a distinct integral coefficient would require an integration instead of a summation. We conclude that if we want to use techniques like DFT in the bubble case, we have to apply them subsequently as described above with the drawback of overconstraining the system of equations.

2.6. Computation of the rational part

The formula to compute the rational part is given in Eq. (2.82). In this section, we describe, how the integral coefficients that enter this equation are computed. Apart from the different tree-level input due to the presence of massive scalars and quarks inside the loop, there are two main algorithmic differences to the cut-constructible part. First, the inclusion of pentagons requires extended kinematics with respect to the external legs. Second, the effective mass μ^2 appears now in the unitarity condition to set the loop momentum on-shell. While it is constant in the pentagon case, it is a free parameter for boxes, triangles and bubbles. It turns out that one can compute the required integral coefficients with an additional Fourier projection over the free parameter μ^2 [67]. How this works is explained in more detail at the example of the box topology. The triangle and bubble case is then treated much shorter since the general concept remains the same.

2.6.1. The rational pentagon contribution

In order to compute the pentagon residue $\bar{e}_{i_1i_2i_3i_4i_5}$ five propagators need to be set on-shell. This generates five equations according to the unitarity constraints $d_{i_1} = d_{i_2} = d_{i_3} = d_{i_4} = d_{i_5} = 0$. Plugging these constraints into Eq. (2.54), the pentagon on-shell loop momentum with the parametrisation of the loop integrand in Eq. (2.5) reads

$$\ell^\mu = V_5^\mu + (\ell \cdot n_\epsilon)\, n_\epsilon^\mu \tag{2.168}$$

where V_5^μ is given by

$$V_5^\mu = -\frac{1}{2}\, q_1^2\, v_1^\mu + \frac{1}{2}\, q_3^2\, v_2^\mu + (q_4^2 - q_3^2)v_3^\mu + (q_5^2 - q_4^2)v_4^\mu. \tag{2.169}$$

Using $V_5 \cdot q_r = \frac{1}{2}q_r^2$, c.f. Eq. (2.56), and $(\ell \cdot n_\epsilon)^2 = -\mu^2$, the five unitarity constraints lead to five identical equations of the form

$$d_{i_r}(\ell) = (\ell - q_r)^2 = V_5^2 - \mu^2 \overset{!}{=} 0. \tag{2.170}$$

We note that the fifth component in the loop momentum in Eq. (2.168) is present only for the purpose of formal manipulations. In practice, it will never appear since we will use the four-dimensional massive vector $\ell_{r[4]}^\mu = V_5^\mu - q_r^\mu$ and the effective mass μ^2 separately. The constant μ-independent pentagon residue is then computed with Eqs. (2.77) and (2.78). In case of a projection onto odd powers in μ as described at the end of section 2.4, the μ-dependent pentagon residue $\bar{e} = e^{(0)} + e^{(1)}\mu$ is evaluated twice with the two independent pentagon masses $\mu = \pm\sqrt{V_5^2}$. The disentanglement gives then

$$e^{(0)} = \frac{1}{2}\Big(\bar{e}(\ell_{[4]}, +\mu) + \bar{e}(\ell_{[4]}, -\mu)\Big), \qquad e^{(1)} = \frac{1}{2\mu}\Big(\bar{e}(\ell_{[4]}, +\mu) - \bar{e}(\ell_{[4]}, -\mu)\Big).$$

Note that $\pm\mu$ enters directly only via propagators or wave functions, the loop momentum is in both cases the same.

Since the pentagon residue $\bar{e}_{i_1i_2i_3i_4i_5}$ enters only as a *constant* subtraction term in the computation of the box, triangle and bubble residue (we ignore the parametrisation in terms of odd powers at the moment), a large numerical value can spoil the numerical accuracy. To estimate the order of magnitude of the numerical value of the pentagon residue, we note that the uniform mass $\mu^2 = V_5^2$ inside the loop is determined from the external kinematics. The absolute value of the pentagon mass

$$\mu_{\text{pen}} = \left|\sqrt{V_5^2}\right| \tag{2.171}$$

sets thus a scale for the pentagon residue $\bar{e}_{i_1i_2i_3i_4i_5}$. In difference to a physical mass scale, this scale depends only on the kinematic configuration. For "well behaved" phase space points, the scale is expected to be around the centre of mass energy. However, inspecting the ingredients that enter the computation of μ_{pen}^2 — basically inverse

Gram determinants — it follows that in collinear phase-space regions this quantity becomes large with respect to other kinematic invariants. Based on the pure knowledge of μ_{pen}, one can estimate with power counting arguments the order of magnitude of the pentagon residue in such phase-space regions. For later use, we illustrate this power counting in a slightly more general way.

If the order of magnitude of the loop momentum components is dominated by a scale Q, and if Q is much larger than all other kinematic invariants, then the order of magnitude of the product of R tree-level amplitudes to reconstruct the mixed quark-gluon integrand as in Eq. (2.77) is to a good approximation

$$A_1 \cdot \ldots \cdot A_R \sim Q^{R-k}. \tag{2.172}$$

R denotes as usual the topology (pentagons, boxes, triangles and bubbles) and k is the number of external quark lines that may enter the loop.[4] In case of a fermion loop, k equals zero and the scaling behaviour is independent of external quarks, like in the pure gluonic case. We consider as an example the above mentioned collinear pentagon case, where $\ell \sim Q \sim \mu_{\text{pen}}$ and therefore for pure gluon amplitudes $\bar{e}_{i_1 i_2 i_3 i_4 i_5} \sim \mu_{\text{pen}}^5$. We will show in the next section that with the above power counting one can adapt a free parameter present in the lower topologies to the size of the pentagon residue leading to a significant improvement of the numerical stability. To this end, we store for every pentagon residue also the corresponding pentagon mass μ_{pen} as a scale for the computations.

2.6.2. The rational box contribution

In order to compute the box residue $\bar{d}_{i_1 i_2 i_3 i_4}(\ell)$ four propagators need to be set on-shell. This generates four equations according to the unitarity constraints $d_{i_1} = d_{i_2} = d_{i_3} = d_{i_4} = 0$. Plugging these constraints into Eq. (2.54), the box on-shell loop momentum reads

$$\ell^\mu = V_4^\mu + (\ell \cdot n_1)\, n_1^\mu + (\ell \cdot n_\epsilon)\, n_\epsilon^\mu = V_4^\mu + \alpha_1\, n_1^\mu + (\ell \cdot n_\epsilon)\, n_\epsilon^\mu \tag{2.173}$$

where $\alpha_1 = \ell \cdot n_1$. V_4^μ and n_1^μ depend only on external quantities and, therefore, are identical to the cut-constructible case, c.f. Eqs. (2.118) and (2.39). Using $V_4 \cdot q_r = \frac{1}{2} q_r^2$, the four unitarity constraints lead to four identical equations of the form

$$d_{i_r}(\ell) = (\ell - q_r)^2 = V_4^2 + \alpha_1^2 - \mu^2 \overset{!}{=} 0. \tag{2.174}$$

---

[4]To understand this relation, recall that the product of R tree-level amplitudes involves at most R off-shell loop propagators less than external legs. The denominators of the off-shell propagators have the form $(\ell - q_i)^2 - \mu^2 = -2\ell \cdot q_i + q_i^2$. Dependent on the flavour content, the loop-momentum dependence in the numerator comes from three-gluon vertices and quark propagators leading to the above scaling.

where $-\mu^2 = (\ell \cdot n_\epsilon)^2$. For an arbitrary choice of μ, this equation has two solutions

$$\alpha_1 = \pm\sqrt{-(V_4^2 - \mu^2)} = \pm\kappa(\mu). \tag{2.175}$$

A difference to the cut-constructible box case in Eq. (2.121) is the μ dependence of α_1. Recalling the general structure of the box integrand in Eq. (2.70)

$$\bar{d}_{i_1 i_2 i_3 i_4}(\ell) = d^{(0)} + d^{(1)}\alpha_1 + d^{(2)}\mu^2 + d^{(3)}\mu^2\alpha_1 + d^{(4)}\mu^4$$

we can disentangle the coefficients $d^{(i)}$ by two subsequent projections: The first projection is over the two values $\alpha_1 = \pm\kappa(\mu)$ exactly like in the cut-constructible case. This allows us to separate the α_1-dependent terms from the α_1-independent terms, i.e. we get the two linear combinations $d^{(0)} + d^{(2)}\mu^2 + d^{(4)}\mu^4$ and $d^{(1)} + d^{(3)}\mu^2$. A subsequent projection over μ^2 then resolves the remaining degeneracies. For the mass μ^2, we choose $(q+1)$ complex numbers equally distributed on a circle of radius r_0^2

$$\mu_m^2 = r_0^2 \exp\left(\frac{2\pi i}{q+1}m\right) \tag{2.176}$$

where $q \in \mathbb{N}$, $q \geq 2$ and $m = 0, \ldots, q$. The choice of the radius r_0^2 — so far a free parameter — will be explained below. The discrete version of the four-dimensional massive loop momentum $\ell_{[4]}^\mu$ then reads

$$\ell_{[4]\sigma,m}^\mu = V_4^\mu + \sigma\,\kappa_m\,n_1^\mu \tag{2.177}$$

with $\sigma = \{\pm 1\}$ and

$$\kappa_m = \sqrt{-\left(V_4^2 - r_0^2 \exp\left(\frac{2\pi i}{q+1}m\right)\right)}. \tag{2.178}$$

The box residue is now computed with Eqs. (2.77) and (2.79). Note that this includes also the subtraction of the pentagon contributions. The explicit disentanglement of the integral coefficients $d^{(i)}$ finally reads

$$d^{(0)} = \frac{1}{2(q+1)} \sum_{m=0}^{q} \sum_{\sigma=\pm} \bar{d}(\ell_{[4]\sigma,m}, \mu_m^2)$$

$$d^{(1)} = \frac{1}{2(q+1)} \sum_{m=0}^{q} \sum_{\sigma=\pm} \frac{\sigma}{\kappa_m}\, \bar{d}(\ell_{[4]\sigma,m}, \mu_m^2)$$

$$d^{(2)} = \frac{1}{2(q+1)} \sum_{m=0}^{q} \sum_{\sigma=\pm} \frac{1}{r_0^2} \exp\left(-\frac{2\pi i}{q+1}m\right) \bar{d}(\ell_{[4]\sigma,m}, \mu_m^2)$$

$$d^{(3)} = \frac{1}{2(q+1)} \sum_{m=0}^{q} \sum_{\sigma=\pm} \frac{1}{r_0^2} \exp\left(-\frac{2\pi i}{q+1}m\right) \frac{\sigma}{\kappa_m}\, \bar{d}(\ell_{[4]\sigma,m}, \mu_m^2)$$

$$d^{(4)} = \frac{1}{2(q+1)} \sum_{m=0}^{q} \sum_{\sigma=\pm} \frac{1}{r_0^4} \exp\left(-\frac{2\pi i}{q+1}2m\right) \bar{d}(\ell_{[4]\sigma,m}, \mu_m^2)$$

$$d^{(\star)} = \frac{1}{2(q+1)} \sum_{m=0}^{q} \sum_{\sigma=\pm} \frac{1}{r_0^4} \exp\left(-\frac{2\pi i}{q+1} 2m\right) \frac{\sigma}{\kappa_m} \bar{d}(\ell_{[4]\sigma,m}, \mu_m^2) \qquad (2.179)$$

The integral coefficient $d^{(\star)}$ corresponds to a rank five tensor structure $\sim \alpha_1 \mu^4$ which is not present in the general decomposition of the box integrand in Eq. (2.70). As a verification of the implementation in NJET we have numerically checked that this coefficient indeed vanishes. The minimal choice for the Fourier parameter is $q = 2$ which corresponds to $2(2+1) = 6$ integrand evaluations. Note that the usage of the DFT overconstrains the system of equations since only five integral coefficients are needed. The only coefficient which explicitly enters the formula for the rational part in Eq. (2.82) is $d^{(4)}$. For the computation of the triangle and bubble coefficients, all coefficients $d^{(0)}, \ldots, d^{(4)}$ are needed. In summary, we see that the rational box coefficients are computed analogously to the cut-constructible case, the major difference is the massive loop momentum and the additional Fourier projection on the mass μ^2. For the disentanglement of the rational box coefficients with odd powers in μ as stated in Eq. (2.111), we must sample additional Fourier modes, now with an explicit choice of μ instead of μ^2:

$$\mu_m = r_0 \exp\left(\frac{2\pi i}{q'+1} m\right) \qquad (2.180)$$

where $q' \in \mathbb{N}$, $q' \geq 4$ and $m = 0, \ldots, q'$.

$$d^{(5)} = \frac{1}{2(q'+1)} \sum_{m=0}^{q'} \sum_{\sigma=\pm} \frac{1}{r_0} \exp\left(-\frac{2\pi i}{q'+1} m\right) \bar{d}(\ell_{[4]\sigma,m}, \mu_m)$$

$$d^{(6)} = \frac{1}{2(q'+1)} \sum_{m=0}^{q'} \sum_{\sigma=\pm} \frac{1}{r_0} \exp\left(-\frac{2\pi i}{q'+1} m\right) \frac{\sigma}{\kappa_m} \bar{d}(\ell_{[4]\sigma,m}, \mu_m)$$

$$d^{(7)} = \frac{1}{2(q'+1)} \sum_{m=0}^{q'} \sum_{\sigma=\pm} \frac{1}{r_0^3} \exp\left(-\frac{2\pi i}{q'+1} 3m\right) \bar{d}(\ell_{[4]\sigma,m}, \mu_m)$$

$$d^{(8)} = \frac{1}{2(q'+1)} \sum_{m=0}^{q'} \sum_{\sigma=\pm} \frac{1}{r_0^3} \exp\left(-\frac{2\pi i}{q'+1} 3m\right) \frac{\sigma}{\kappa_m} \bar{d}(\ell_{[4]\sigma,m}, \mu_m) \qquad (2.181)$$

Note that in the presence of odd powers in μ, the disentanglement of the even coefficients $d^{(0)}, \ldots, d^{(4)}$ must be performed with the increased number of Fourier modes q' as well *and* a modified exponential basis function with the substitution $m \to 2m$. Furthermore, the pentagons are subtracted using the odd, μ-dependent parametrisation of $\bar{e}(\mu)$ given in Eq. (2.110).

We conclude this section with a remark on the choice of the Fourier radius r_0. An analytic computation is independent of r_0. However, numerically, we have to choose a suitable scale, since r_0 is a dimensionful quantity. This can have drastic effects on the numerical accuracy. It turns out that a dynamical radius which is determined for every box coefficient separately stabilises the result, a procedure which we will describe in the following.

The bottleneck of the numerical accuracy is the subtraction of the constant pentagon contributions to get the box integrand which schematically reads

$$\bar{d}(\ell, \mu^2) = A_1 \cdot A_2 \cdot A_3 \cdot A_4 - \sum_{i \notin C_4} \frac{e_i^{(0)}}{-2\ell q_i + q_i^2} \, . \tag{2.182}$$

$A_1 \cdot A_2 \cdot A_3 \cdot A_4$ is the product of tree amplitudes as in Eq. (2.77), i labels topologies with common propagators and q_i is the momentum parametrisation for the propagator in question. Further we brought the inverse propagator which divides the pentagon residue into the form $(\ell - q_i)^2 - \mu^2 = -2\ell q_i + q_i^2$ to make the linear dependence on the loop momentum explicit. It follows from Eq. (2.182) that a large numerical value of $e_i^{(0)}$ can lead to big cancellations and, therefore, can become numerically unstable. The most problematic situation is encountered if a collinearity appears only in the pentagon topology leading to a large pentagon residue and large pentagon mass μ_{pen} while the box outflow momenta are all non-collinear. The natural scale of the box topology V_4^2 is then much smaller than the natural scale of the pentagons μ_{pen}^2, i.e. $V_4^2 \ll \mu_{\text{pen}}^2$. The product of tree-level amplitudes will be large as well (the collinearities enhance the external propagators), but after having subtracted the large pentagon contribution, the box residue $\bar{d}(\ell, \mu^2)$ can become small and much less accurate. Since the pentagon residue is constant one cannot entirely circumvent this problem. A significant improvement can be achieved, however, if we choose a large value for the radius r_0 such that the components of the loop momentum are dominated by r_0. In a rather sloppy notation $\ell \sim r_0$. This has two consequences: First it dampens the subtraction term and second it increases simultaneously the contributions of the tree-level product according to Eq. (2.172). Hence, after subtracting the pentagon contribution the remaining terms are still of the same order of magnitude. Possible cancellations will then occur first after solving the system of equations and not already in advance.

A choice for the radius r_0 which incorporates the required features is simply the *largest absolute value of the pentagon mass* $\mu_{\text{pen},i}$ whose corresponding pentagon coefficient $e_i^{(0)}$ appears in the subtraction in Eq. (2.182):

$$r_0 = \mu_{\text{pen,max}} = \max_i (|\mu_{\text{pen},i}|). \tag{2.183}$$

According to Eq. (2.172), the size of the pentagon residue can be estimated as $\bar{e}_{i_1 i_2 i_3 i_4 i_5} \sim \mu_{\text{pen}}^5$.[5] On the other hand, with this choice of the radius, the box loop momentum is dominated by $r_0 = \mu_{\text{pen,max}}$ as well. Dividing by the inverse propagator hence dampens the largest subtraction term down to an order $\mu_{\text{pen,max}}^4$ contribution, any other pentagon subtraction term will be smaller. Controversially, the product of four tree-level amplitudes is increased according to $\sim r_0^4 = \mu_{\text{pen,max}}^4$, c.f. Eq. (2.172). Hence, one gains that all terms are of the same order of magnitude. Since all box outflow momenta are part of the pentagon outflow momenta, the dynamical adaption of the radius works

---

[5] We focus here on the gluonic case, the quark case works analogously but does not bring fundamentally new insights.

also if the collinearities appear both in the boxes and pentagons. For uncritical phase-space points the dynamical radius settles down roughly around the centre of mass energy, a natural scale present in massless QCD.

2.6.3. The rational triangle contribution

The strategy for the rational triangle contribution is conceptually similar to the rational box case. The external kinematics involving V_3^μ, n_1^μ and n_2^μ is the same as for the cut-constructible triangles. Setting three propagators on the mass shell with five-dimensional loop momentum leads to the unitarity constraint

$$\alpha_1^2 + \alpha_2^2 = -(V_3^2 - \mu^2) \equiv \kappa^2(\mu^2). \tag{2.184}$$

The difference to the unitarity constraint of the cut-constructible triangles is that κ does now depend on the additional free parameter μ^2. The strategy is as in the box case to do two projections, one with respect to α_1 and α_2 exactly as in the cut-constructible triangle case, and the second one with respect to μ^2 as in the rational box case. We choose μ^2 on a circle of radius r_0^2, precisely as in Eq. (2.176). The value of r_0 is again determined dynamically adapting its size to the largest pentagon subtraction term in exactly the same manor as described in the rational box case. Making sure that $r_0^2 \neq V_3^2$, then $\kappa^2(\mu^2) \neq 0$ for all sampled points. This means that we need to consider only the "massive" triangle parametrisation in Eqs. (2.137) and (2.138), c.f. the detailed discussion in section 2.5.2. We denote the discrete μ^2-dependence of κ with an additional index m:

$$\kappa_m = \sqrt{-\left(V_3^2 - r_0^2 \exp\left(\frac{2\pi i}{q+1}m\right)\right)}. \tag{2.185}$$

The discrete version of the four-dimensional massive loop momentum then reads

$$\ell_{[4]j,m}^\mu = V_3^\mu + \kappa_m \sin\left(\frac{2\pi}{2p+1}j\right) n_1^\mu + \kappa_m \cos\left(\frac{2\pi}{2p+1}j\right) n_2^\mu. \tag{2.186}$$

The computation of the triangle residue $\bar{c}(\ell_{[4]j,m}, \mu_m^2)$ proceeds with Eqs. (2.77) and (2.80). Note that one has to subtract both contributions from pentagons and boxes. Comparing the general form of the triangle integrand in Eq. (2.71) with the purely four-dimensional one in Eq. (2.130), we see that we have to include three additional terms $\sim \alpha_1\mu^2$, $\alpha_2\mu^2$ and $\sim \mu^2$. For the disentanglement of the coefficients $c^{(i)}$ we use first the projection from the cut-constructible triangles in Eqs. (2.146) with a fixed value of μ_m^2 (fixed index m). This is made explicit in the following with the notation $P[i] \to P_m[i]$.

This is followed by the projection onto the effective mass μ_m^2.

$$c^{(0)} = \frac{1}{q+1} \sum_{m=0}^{q} P_m[0] \equiv \frac{1}{q+1} \sum_{m=0}^{q} \frac{1}{2p+1} \sum_{j=-p}^{p} \bar{c}(\ell_{[4]j,m}, \mu_m^2)$$

$$c^{(1)} = \frac{1}{q+1} \sum_{m=0}^{q} \frac{1}{\kappa_m} (P_m[1] - P_m[6]) \qquad c^{(2)} = \frac{1}{q+1} \sum_{m=0}^{q} \frac{1}{\kappa_m} (P_m[2] + P_m[5])$$

$$c^{(3)} = \frac{1}{q+1} \sum_{m=0}^{q} \frac{-1}{\kappa_m^2} P_m[3] \qquad c^{(4)} = \frac{1}{q+1} \sum_{m=0}^{q} \frac{2}{\kappa_m^2} P_m[4]$$

$$c^{(5)} = \frac{1}{q+1} \sum_{m=0}^{q} \frac{-4}{\kappa_m^3} P_m[5] \qquad c^{(6)} = \frac{1}{q+1} \sum_{m=0}^{q} \frac{4}{\kappa_m^3} P_m[6]$$

$$c^{(7)} = \frac{1}{q+1} \sum_{m=0}^{q} \frac{1}{r_0^2} \exp\left(-\frac{2\pi i}{q+1} m\right) \frac{1}{\kappa_m} (P_m[1] - P_m[6])$$

$$c^{(8)} = \frac{1}{q+1} \sum_{m=0}^{q} \frac{1}{r_0^2} \exp\left(-\frac{2\pi i}{q+1} m\right) \frac{1}{\kappa_m} (P_m[2] + P_m[5])$$

$$c^{(9)} = \frac{1}{q+1} \sum_{m=0}^{q} \frac{1}{r_0^2} \exp\left(-\frac{2\pi i}{q+1} m\right) P_m[0] \tag{2.187}$$

For the coefficient $c^{(0)}$ we have written out explicitly the projection $P[0]$ to make the general idea clear. For the computation of the rational part, only $c^{(9)}$ enters Eq. (2.82). However, all other coefficients are also explicitly needed as subtraction terms for the evaluation of the bubble integrand. In NJET we found empirically that with three Fourier modes ($q = 2$), the inversion of the system of equations becomes stable. For higher modes, there is basically no improvement. The price to pay is that the integrand is evaluated $3 \times 7 = 21$ times which is roughly double the number of the ten independent integral coefficients. Although the DFP was among those methods that we tested to solve the system of equations the most stable one, searching for faster and numerically equally stable techniques is in this case recommended. The investigation of this point is postponed for future studies. The disentanglement of possible additional odd powers in μ as described in Eq. (2.112) can be achieved in full analogy to the box case described in the previous section.

2.6.4. The rational bubble contribution

In the rational bubble case, a simplification occurs because from the ten integral coefficients in Eq. (2.72) only the coefficient $b^{(9)}$ is needed. The external kinematics involving V_2^μ, n_1^μ, n_2^μ and n_3^μ are the same as for the cut-constructible bubbles. Setting two propagators on the mass-shell with five-dimensional loop momentum leads to the unitarity constraint

$$\alpha_1^2 + \alpha_2^2 + \alpha_3^2 = -(V_2^2 - \mu^2) \equiv \kappa^2(\mu^2). \tag{2.188}$$

The strategy is again to employ the Fourier projection from the cut-constructible bubbles leading to $b^{(0)} + \mu^2 b^{(9)}$, followed by a Fourier projection over the free parameter μ^2 which projects out the coefficient $b^{(9)}$. We employ the same parametrisation for μ^2 as in Eq. (2.176), especially the Fourier radius r_0 is again adapted to the largest pentagon subtraction term as described in section 2.6.2 on the rational box coefficients. Recall from Eq. (2.81) that the computation of the bubble residues involves the subtraction of triangle, box and pentagon contributions. The discretised version of κ in the bubble case simply reads

$$\kappa_m = \sqrt{-\left(V_2^2 - r_0^2 \exp\left(\frac{2\pi i}{q+1}m\right)\right)}. \tag{2.189}$$

with $q \in \mathbb{N}, q \geq 1$ and $m = 0, \ldots, q$. Correspondingly, one finds for the four-dimensional massive loop momentum

$$\ell_{[4]j_x, j_y, m}^{\mu} = V_2^{\mu} + \kappa_m \sin\left(\frac{2\pi}{2p_x+1}j_x\right)\cos\left(\frac{2\pi}{2p_y+1}j_y\right)n_1^{\mu} \tag{2.190}$$

$$+ \kappa_m \sin\left(\frac{2\pi}{2p_x+1}j_x\right)\sin\left(\frac{2\pi}{2p_y+1}j_y\right)n_2^{\mu} + \kappa_m \cos\left(\frac{2\pi}{2p_x+1}j_x\right)n_3^{\mu}.$$

The range of the integers j_x, j_y, p_x and p_y is defined in section 2.5.3. With these ingredients at hand, the explicit formula for the disentanglement reads

$$b^{(9)} = \frac{1}{q+1}\frac{1}{(2p_x+1)(2p_y+1)}\sum_{m=0}^{q}\sum_{j_x=-p_x}^{p_x}\sum_{j_y=-p_y}^{p_y}\bar{b}(\ell_{[4]j_x,j_y,m}, \mu_m^2) \times$$

$$\times \frac{1}{r_0^2}\exp\left(-\frac{2\pi i}{q+1}m\right)\left(1 - \frac{2}{3}\cos\left(\frac{2\pi}{2p_x+1}2j_x\right)\right). \tag{2.191}$$

Note that this solution involves three independent Fourier projections. The minimal choice of the Fourier parameter is $q = 1$ which corresponds to two Fourier modes. As discussed in chapter 2.5.3, the number of Fourier modes for the cut-constructible bubbles is 15. The full disentanglement requires therefore $2 \times 15 = 30$ integrand evaluations. Compared with the 10 independent coefficients, the system of equations is overconstrained by a factor of three. On the other hand, the computation of the bubble coefficient is from the numerical point of view the most challenging one because it includes an intricate subtraction procedure involving triangles, boxes and pentagons. From those methods that we tested to solve the system of equations, the discrete Fourier transformation led to the best numerical accuracy. In appendix D.2, an alternative disentanglement is proposed which requires only eight integrand evaluations to get the full bubble coefficient $b^{(9)}$. From the performance point of view this is almost a factor of four of improvement. Since the numerical accuracy is not tested yet, we cannot make any judgement on the goodness of such an approach.

2.7. Tree-level amplitudes for the unitarity cuts

The tree-level amplitudes that enter unitarity cuts of one-loop primitive amplitudes are basically the tree-level primitive amplitudes described in chapter 1. The essential difference is that always two adjacent legs are now part of the loop and have, therefore, in general complex momenta. In the case of trees for the rational part, the loop legs and propagators are massive and involve scalars instead of gluons. In the following, we will call tree-level amplitudes that enter unitarity cuts *subamplitudes*.

Let all quantities be given that uniquely specify an n-point one-loop primitive amplitude: a list of external momenta $\mathcal{P} = [p_1, \ldots, p_n] \equiv [1, \ldots, n]$, flavours $\mathcal{F}_{\text{ext}} = [f_1, \ldots, f_n]$, helicities $\mathcal{H}_{\text{ext}} = [h_1, \ldots, h_n]$ and the list of the propagators' flavour in the parent diagram $\mathcal{F}_{\text{par}} = [F_1, \ldots, F_n]$. We will use the momentum list $\mathcal{P}$ also as permutation, it should be clear from the context what is actually meant. A subamplitude within an n-point primitive is then uniquely identified with the choice of:

1. A distinct topology $C_R = \{i_1, \ldots, i_R\}$

2. A position within the topology $i_r \in C_R$, c.f. the definitions in section 2.1.1.

3. A specifier $w \in \{0, 1\}$ to indicate whether one is dealing with the cut-constructible ($w = 0$) or the rational part ($w = 1$).

4. The Fourier modes x which parametrise the on-shell loop momentum, c.f. the explicit parametrisation in Eqs. (2.122), (2.144), (2.152), (2.165), 2.177), (2.186) and (2.190).

An m-point subamplitude, therefore, explicitly reads:

$$A_m^{\text{cut-tree}}(C_R, i_r, x, w) = A_m^{\text{cut-tree}}(\{-\ell_r^{C_R;x;w}, \bar{F}_{i_r}\}, \{p_{i_r}, f_{i_r}\}, \ldots, \tag{2.192}$$
$$\{p_{i_{(r+1)}-1}, f_{i_{(r+1)}-1}\}, \{\ell_{r+1}^{C_R;x;w}, F_{i_{(r+1)}}\}) \, .$$

where the multiplicity m is determined as $m = 2 + [(n + i_{(r+1)} - i_r) \bmod n]$ and $\bar{F}_{i_r}$ is the anti-particle of F_{i_r}. For a tree contributing to the rational part ($w = 1$), every gluon in $\mathcal{F}_{\text{par}}$ turns into a scalar. The helicity labels are suppressed in order not to clutter up the notation. Note the explicit dependence of the loop momentum on the full topology $C_R = \{i_1, \ldots, i_R\}$. This is because the on-shell loop momentum is constructed with all elements of the kinematic configuration $\mathcal{K}_R = \{k_1, \ldots, k_R\}$ induced by C_R.

There is one additional subtlety to consider when computing mixed primitives $A^{[m]}$ with external quark lines: In a topology $C_R = \{i_1, \ldots, i_R\}$ where the first and the last on-shell propagator F_{i_1} and F_{i_R} have the same quark flavour (recall that for mixed primitives F_1 is always a gluon, c.f. the discussion in section 1.3), the subamplitude which is classified by i_R looks like an equal-flavour quark amplitude: It involves two external quark legs and two loop quark legs which — seen from from the level of the one-loop amplitude — belong to one single quark line. At the subamplitude level, however, they must be treated as if they were different. Otherwise, one would include

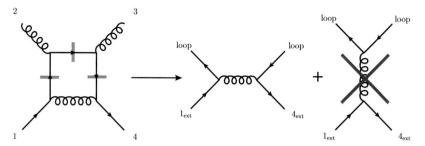

Figure 2.5.: Faked like-flavour quark subamplitude in a triple cut with topology $C_3 = \{2, 3, 4\}$: The subamplitude characterised by $i_3 = 4$ involves two quark lines of the same flavour which must be treated as different ones to exclude contributions from a closed fermion loop.

contributions that belong to a closed quark loop. The situation is exemplified in Fig. 2.5 for a triple cut of a four-point primitive $A^{[m]}(\bar{q}, g, g, q)$ with topology $C_3 = \{2, 3, 4\}$. Such situations can occur only in subleading colour primitive amplitudes with non-vanishing quark–anti-quark separation.

Since all tree-level amplitudes are computed with Berends-Giele recursion, the contributing off-shell currents that involve exclusively external legs remain unchanged for different unitarity cuts as long as the external helicities and the external legs' order remain unchanged as well. In appendix B.1 it is shown that in this case a suitable cache system can improve the asymptotic behaviour of recursive tree-level evaluations from the well known $O(n^4)$ scaling down to an effective $O(n^3)$ behaviour for the mixed quark–gluon case, and even further down to $O(n^2)$ for the closed quark loop case. The main idea is to compute in an initialisation phase all possible external off-shell currents that may appear within one distinct primitive amplitude, and then to reuse them in the computation of subamplitudes for individual unitarity cuts. How off-shell recursion is realised in NJET in its most general form including quarks, gluons and the scalars for the rational part is described in appendix B.2.

Cache system for tree-level subamplitudes: To compute all integral coefficients we must iterate through all possible topologies C_R, once for the cut-constructible part ($w = 0$) and once for the rational part ($w = 1$), at any time with x integrand evaluations for the different Fourier modes. For any combination of C_R, w and x, the on-shell loop momenta are different. Within one primitive amplitude with fixed order of external legs an helicities, every subamplitude appears exactly once. (We assume that the sum over internal helicities is optimised such that every helicity configuration is computed only once.) As soon as we compute full-colour helicity summed squared matrix elements, primitives with different $\mathcal{P}$, $\mathcal{F}_{\text{ext}}$, $\mathcal{H}_{\text{ext}}$ and $\mathcal{F}_{\text{par}}$ contribute. Those subamplitudes that agree in momenta, flavour and helicities of both external legs and internal legs can be reused for different primitive amplitudes. An example is illus-

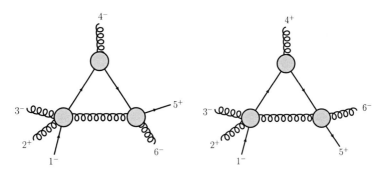

Figure 2.6.: Triple cut for two primitives with different helicity and ordering of external legs. The subamplitudes $A_5^{\text{cut-tree}}(\ell_{1;g}^{\pm}, 1_q^-, 2_g^+, 3_g^-, \ell_{4;\bar{q}}^{\pm})$ are the same for both primitives.

trated in Figure 2.6 for a triangle cut of a six-point amplitude. Flipping the helicity of the fourth external gluon and permuting the fifth and sixth external leg leaves the five point subamplitudes $A_5^{\text{cut-tree}}(\ell_{1;g}^{\pm}, 1_q^-, 2_g^+, 3_g^-, \ell_{4;\bar{q}}^{\pm})$ unchanged.

As a key to access individual subamplitudes $A_m^{\text{cut-tree}}(C_R, i_r, x, w)$ and to reuse them through a cache, we employ a triplet of non-negative integer numbers $X = (X_1, X_2, X_3)$. X_1 incorporates information of the subamplitude which depends on the whole topology $C_R = \{i_1, \ldots, i_R\}$. X_2 contains information that is specific for the individual subamplitude like external permutations and helicities describable with i_r and $i_{(r+1)} - 1$ given $\mathcal{P}$ and $\mathcal{H}_{\text{ext}}$. In addition, it includes also the Fourier modes x and the rational part w. X_3 is dedicated to the flavour information. At the programming level in C++, one can define order relations to put all triplets X into lexicographic order. The container class <map> from the C++ standard template library is then able to construct an internal binary tree of subamplitudes whose elements are accessed with the triplets X, i.e. X works like a single index. Since we have to sum over internal helicities, we store always a block of four amplitudes belonging to the four helicity combinations $(+, +)$, $(+, -)$, $(-, +)$ and $(-, -)$ of the two partons connecting the amplitude to the loop. We explain now how to construct the triplets X and finally comment on possible improvements.

As already anticipated, the explicit dependence on the topology C_R in $A_m^{\text{cut-tree}}(C_R, i_r, x, w)$ is due to the kinematic configuration $\mathcal{K}_R = \{k_1, \ldots, k_R\}$ from which the loop momentum is constructed. Since any k_r is a momentum sum of external particles given in Eq. (2.1), two kinematic configurations of different permutations $\mathcal{K}_R(\mathcal{P})$ and $\mathcal{K}_R(\mathcal{P}')$ agree if their momentum sums agree. In order to compare kinematic configurations with each other, we divide the permutation list $\mathcal{P}$ for a given topology C_R into R sublists $[\mathcal{P}_1, \ldots, \mathcal{P}_R]$ where an individual sublist $\mathcal{P}_r$ reads

$$\mathcal{P}_r = [p_{i_r}, \ldots, p_{i_{(r+1)}-1}]. \tag{2.193}$$

As usual, the indices are cyclic in n and R as described in section 2.1.1. In a next step, every sublist is brought in total ordering starting with the smallest element of $\mathcal{P}_r$ and ending with the largest one. For a given C_R, we thus have a partially ordered new permutation list

$$\mathcal{P}_{\mathrm{ord}}(\mathcal{P}, C_R) = [\mathcal{P}_1^{\mathrm{ord}}, \dots, \mathcal{P}_R^{\mathrm{ord}}]. \tag{2.194}$$

Two kinematic configurations are thus equal if both C_R and $\mathcal{P}_{\mathrm{ord}}$ agree with each other. An example for this procedure is the permutation $\mathcal{P}_6 = [1, 3, 2, 6, 5, 4]$ with the triangle topology $C_3 = \{2, 4, 6\}$. The three sublists are $\mathcal{P}_1 = [3, 2]$, $\mathcal{P}_2 = [6, 5]$, $\mathcal{P}_3 = [4, 1]$ which lead to the ordered list $\mathcal{P}_{\mathrm{ord}} = [2, 3, 5, 6, 1, 4]$. To assign every kinematic configuration $\mathcal{K}_R$ a unique integer number $\mathcal{N}_{\mathcal{K}_R}(\mathcal{P}, C_R) \equiv X_1$, we assign both C_R and $\mathcal{P}_{\mathrm{ord}}$ a separate number and combine them appropriately. As a side effect, we use $\mathcal{N}_{\mathcal{K}_R}(\mathcal{P}, C_R)$ to store also the scalar master integrals and the basis vectors V_R^μ and n_i^μ of the NV basis since they depend exclusively on the kinematic configuration $\mathcal{K}_R$.

We employ a factoradic number system which maps each of the $N!$ permutations $\mathcal{P} = [p_1, \dots, p_n]$ with $p_i \in \{1, \dots, n\}$ to a unique number $\mathcal{N}_{\mathrm{perm}}(\mathcal{P}) \in \{0, \dots, n! - 1\}$. The assignment reads

$$\mathcal{N}_{\mathrm{perm}}(\mathcal{P}) = \sum_{i=0}^{n-2} (n-1-i)! \sum_{j=i+1}^{n-1} \Theta(p_i - p_j) \tag{2.195}$$

where the theta function is defined as

$$\Theta(p_i - p_j) = \begin{cases} 1 & \text{for} \quad p_i > p_j \\ 0 & \text{for} \quad p_i \le p_j \end{cases}. \tag{2.196}$$

In order to assign also any topology $C_R = \{i_1, \dots, i_R\}$ a unique number we employ the formula

$$\mathcal{N}_{\mathrm{topo}}(C_R) = \sum_{r=2}^{R-1} \binom{n}{r} + \sum_{r=1}^{R} \binom{i_r - 1}{r}. \tag{2.197}$$

The first sum counts the total number of all possible lower topologies with smaller R, the second sum enumerates topologies for fixed R.[6] Its range is $\mathcal{N}_{\mathrm{topo}}(C_R) \in \{0, \dots, \mathcal{N}_{\mathrm{topo}}^{\mathrm{tot}} - 1\}$ with

$$\mathcal{N}_{\mathrm{topo}}^{\mathrm{tot}} = \sum_{r=2}^{5} \binom{n}{r} \tag{2.198}$$

---

[6]We ignore in the general derivation that this counting includes also on-shell bubbles which are zero in the massless case. In NJET this is taken into account to save memory.

being the total number of topologies. Hence, the function

$$X_1 \equiv \mathcal{N}_{\mathcal{K}_R}(\mathcal{P}, C_R) = \mathcal{N}_{\text{topo}}(C_R) + \mathcal{N}_{\text{topo}}^{\text{tot}} \cdot \mathcal{N}_{\text{perm}}\Big(\mathcal{P}_{\text{ord}}(\mathcal{P}, C_R)\Big). \tag{2.199}$$

assigns every independent kinematic configuration $\mathcal{K}_R$ a unique integer number.

The mapping of external quantities from $A_m^{\text{cut-tree}}(C_R, i_r, x, w)$ to X_2 and X_3 proceeds via bitwise operations. We choose to represent every external particle of the permutation list with four bits in a binary representation. This allows in principle to represent the numbers $0, \ldots, 15 < 2^4$. The external helicities are mapped from $(-, +) \to (0, 1)$ and are represented by a single bit. First, we initialise $X_{2,(0)} = 0$, i.e. we set all bits equal to zero, the additional subscript denotes an iteration step. Subsequently, we add $X_{2,(1)} = X_{2,(0)} + p_{i_r}$, followed by shifting all bits one digit to the left such that also h_{i_r} is added $X_{2,(3)} = X_{2,(2)} + h_{i_r}$. This is now followed by shifting all bits four digits to the left such that one can add $X_{2,(4)} = X_{2,(3)} + p_{i_r+1}$, followed again by the helicity h_{i_r+1} and so on. This is iterated until one reaches position $i_{(r+1)} - 1$. The result is a unique bit pattern which describes a certain permutation and helicity configuration of the subamplitude. These cut dependent operations are followed by three constant left shifts: one digit for w to indicate the rational or cut-constructible part, seven digits for the Fourier modes (this represents an upper limit also for higher Fourier modes) and one digit to indicate whether we used a scaling test to check the numerical accuracy (we will explain this test in detail in section 3.1). We interpret the final bitpattern as the binary representation of an integer number which was the requirement of the triplet $X = (X_1, X_2, X_3)$. Choosing the X_i of data type `unsigned long` with a data type model where `long` has 64 bits, this allows to cache subamplitudes from primitives with up to 13 external legs: $(13-2)*(4+1)+1+7+1 = 64$. The flavour structure is implemented in X_3 in a similar way. The 13 external legs permit 6 quark–anti-quark pairs with indices $\{-1, +1\}, \ldots, \{-6, +6\}$ (c.f. appendix B.2 for the definition of the quark index system). If the cache system is switched on, the flavour for the closed quark loop is automatically set to ± 7. Adding to all flavour indices $+7$ (including the gluon with index 0), we have only non-negative flavour numbers with range $0, \ldots, 14 < 2^4$ which can again be represented by four bits. The assignment proceeds as for X_2. A constant left shift after the cut dependent shifts includes also the loop flavour.

A serious disadvantage of the presented cache system is that all possible subamplitudes are unsorted in the same container. Although the average access to individual objects in a binary tree scales as $\log(n)$ with n being the number of cached objects [138], the total amount of subamplitudes is so large that both searching and the internal management of the binary tree becomes a significant overhead — recall that the computation of a full colour amplitude scales factorially with the number of primitives. A possible improvement would be to sub-divide the cache such that amplitudes from common classes of topologies have their own cache, i.e. a separate cache for the cut-constructible boxes, the rational pentagons etc. The effect of the cache system on the runtime performance is discussed in section 3.2.

3. Numerical computation of one-loop QCD amplitudes

Both the tree-level and one-loop methods described in the previous chapters 1 and 2 have been implemented in a numerical C++ program publicly available as the library NJET [80]. It is based on our previously published program NGLUON [77] which is dedicated to the computation of primitive amplitudes in pure gauge theory with arbitrarily many external legs. We note that NGLUON was the first publicly available program allowing for computing gluon primitive amplitudes with arbitrary multiplicity and helicity configuration. NGLUON has been extended to compute also primitive amplitudes with an arbitrary number of massless external quark lines [78], and full colour and helicity summed interference terms of virtual corrections and Born amplitudes for all processes contributing to 2-jet, 3-jet, 4-jet and 5-jet production in massless QCD [79]. The colour summed helicity amplitudes are implemented in NJET as described in section 1.4 such that only independent primitive amplitudes are evaluated. Finally, the helicity summation is performed by calculating the colour summed helicity amplitudes. In this chapter, we will describe the validation of the implementation, the numerical accuracy of the amplitude evaluation and the runtime performance. The program is equipped with the extended precision package QD [139] which allows to compute all integral coefficients in quadruple precision. For the evaluation of the scalar one-loop integrals we use the external library FF/QCDLOOP [117, 118]. The explicit usage of the program is explained in Ref. [80] and shall not be the subject of this work. Some of the results and arguments that are presented in the following are based on our publications on NGLUON [77] and NJET [80] and on the conference proceedings from a talk which the author gave at ACAT 2011 [78].

3.1. Validation and numerical accuracy

The correctness of the numerical implementation has been verified by a series of highly non-trivial tests, both at the level of primitive amplitudes and at the level of colour summed interferences between Born and virtual corrections. A very strong check to validate the entire cut-constructible part is the correct reproduction of the universal pole structure of the IR and UV poles presented analytically in section 2.1.4. Expanding the amplitude in terms of the dimensional regulator ϵ up to order ϵ^0

$$A = \frac{1}{\epsilon^2} a_{(-2)} + \frac{1}{\epsilon^1} a_{(-1)} + a_{(0)} + O(\epsilon) \tag{3.1}$$

where $a_{(0)}$ denotes the finite part of the amplitude and $a_{(-1)}$ respectively $a_{(-2)}$ the co-efficient of the corresponding poles in ϵ, NJET allows the numerical evaluation of the coefficients $a_{(-1)}$, $a_{(-2)}$ and $a_{(0)}$. Since only the four-dimensional integral coefficients from section 2.5 contribute to $a_{(-1)}$ and $a_{(-2)}$, the poles are evaluated with almost no computational effort, once the finite part $a_{(0)}$ is calculated. The IR poles occur only in certain scalar triangle and box integrals. The correct reproduction of the IR poles serves thus as an explicit test only for a subset of the triangle and box coefficients. As far as the UV poles are concerned this test checks a linear combination of the bubble integral coefficients. Yet, this is an implicit test of the entire cut-constructible part be-cause first, the triangle and box coefficients enter during the computation of the bubble coefficients as subtraction terms, and because second, all bubble coefficients contribute to the UV poles. For all processes that we have tested we find perfect agreement with the analytic formulae given explicitly in Eqs. (2.14) and (2.16) for arbitrary primitive amplitudes, and in Eq. (2.10) for the colour summed interference between Born and one-loop amplitude (checked for all channels of two-jet, three-jet, four-jet and five-jet production). Since all four-dimensional integral coefficients are computed with the same algorithm described in section 2.5, the correct pole structure represents a very strong check for the whole cut-constructible part. The rational part, however, cannot be tested in this way. For this reason, we employed analytic results for the rational part of specific helicity configurations which hold for arbitrary multiplicity as analytic cross check. For the pure gluon primitive amplitudes, this is the all-n-MHV-formula for the helicity configuration $- - + \ldots +$ [140]. We checked the rational part obtained numerically from NGLUON with this formula up to multiplicities of $n = 20$ and found agreement. Since the implementation does not depend on the helicities of the external particles, this gives us very strong confidence that the implementation is also correct for arbitrary gluon helicities. Further, we used the IR- and UV-finite all-n-formula for primitives with a single quark–anti-quark pair and $(n - 2)$ positive helicity glu-ons $A_n^{[m/f]}(\bar{q}_1^-, g_2^+, \ldots, q_i^+, \ldots, g_n^+)$ of Ref. [141] as analytical cross-check. We checked all possible quark–anti-quark separations, for multiplicities up to $n = 20$ both for the mixed quark-gluon loop and the closed quark loop. Again, we found agreement. Since the same algorithm is also applied in the multi-fermion case, this gives us very strong confidence that the rational terms are correct for any quark and helicity configuration. Together with the internal pole check for the cut-constructible part, we are convinced that the presented implementation is correct.

To estimate the numerical accuracy of the implementation, we analysed the accu-racy for a large number of phase space points. In a Monte Carlo integration, not all regions in phase space contribute with the same weight to the cross section. For exam-ple, collinear momentum configurations are often enhanced because of the numerically larger value of the amplitudes. These configurations are in general numerically less stable due to the presence of small inverse Gram determinants at intermediate steps. To mimic this behaviour, we employed a sequential splitting algorithm as described in Ref. [142] to generate sample phase space points. We have investigated the properties of this phase space generator in Ref. [113]. Compared with a generator for a flat phase space as for example RAMBO [143], sequential splitting favours collinear momentum configurations. It gives, thus, more realistic estimates of the accuracy although some accuracy distributions may look worse than those generated from a flat phase space.

A way to access the numerical uncertainty of a quantity A due to rounding errors or numerical cancellations is to compute it twice with different methods and, then, to compare the two results A_1 and A_2 with each other. We define the relative accuracy to be

$$\delta(A) = \log_{10}\left(\left|2\frac{A_1 - A_2}{A_1 + A_2}\right|\right). \tag{3.2}$$

Hence, $-\delta$ estimates the number of common digits between the two quantities. We may, for example, employ A_1 and A_2 to compare a purely numerical quantity obtained from NJET with a numerically evaluated analytical formula. Estimating the accuracy via analytical formulae is first of all limited to the availability of analytic results. Hence, to make a statement on an arbitrary phase space point for an arbitrary amplitude which is computable with NJET — for example a 20 gluon primitive amplitude with alternating helicity configuration — the accuracy must be determined internally, independent of additional external input. One can, for example, employ the analytically known universal pole structure to compare with the numerically computed poles of the current evaluation: The size of possible deviations is indeed an indication for numerical instabilities of the finite part. However, as argued before, this will not test the rational part at all. Another option is to do the discrete Fourier transformation that we discussed in sections 2.5 and 2.6 with additional Fourier modes to project on tensor structures which necessarily vanish and to relate the goodness of such a "manually" computed zero to the accuracy of the non-vanishing integral coefficients. An example of such an additional Fourier mode has been described in the rational box case in section 2.6.2 where we always projected out an additional vanishing coefficient of a rank five tensor structure. To some extent, this approach gives hints on numerical instabilities of individual integral coefficients, though a quantitative and rigorous statement on the number of valid digits could not be established and we did not follow this approach any further. Note that even if we knew the numerical error of an individual integral coefficient, a statement on the overall accuracy of the amplitude can first be made if we know the error of every single integral coefficient contributing to the primitive amplitude and, in addition, also the numerical value (and the error) of the scalar master integrals. For example, it happens that some integral coefficients are entirely wrong, yet, other larger terms with at least some valid digits can still dominate the amplitude and lead to an acceptable result.[1]

As a solution of these problems, we have proposed in Ref. [77] the so called *scaling*

---

[1]We note that the scalar integrals computed with the external library QCDLOOP are only available in double precision which limits the final result of the cut-constructible part and, therefore, in general, of the full amplitude to at most 16 decimal digits even if extended precision is used. For phenomenological applications, this is not a problem because the calculation of the integral coefficients with possibly large subtraction terms will always dominate the overall accuracy. It is thus sufficient to use extended precision in the reduction process alone.

test: It checks the amplitude as a whole, is independent of additional external input, and, furthermore, is very simple to apply. The basic idea is that if the dimensionful parameters which the amplitude depends on are rescaled with a distinct factor x, the amplitude — being itself in general a dimensionful quantity — changes its value with integer powers in x. Physically, this is nothing else than computing the amplitude with quantities expressed in different units (for example GeV and TeV corresponding to $x = 1000$). In our case, the dimensionful input parameters are the external momenta and the renormalisation scale. A simple power counting analysis dictates the relation between scaled and unscaled n-point amplitudes to be $A_n(xp, x\mu_R) = x^{4-n} A_n(p, \mu_R)$ where $xp = \{xp_1, \ldots, xp_n\}$. Every amplitude is now computed twice, first with a momentum configuration p and μ_R, and another time with xp and $x\mu_R$ such that the relative uncertainty can be estimated with help of Eq. (3.2) as

$$\delta(A(p, \mu_R)) = \log_{10} \left(\left| 2 \frac{A(p, \mu_R) - x^{n-4} A(xp, x\mu_R)}{A(p, \mu_R) + x^{n-4} A(xp, x\mu_R)} \right| \right). \tag{3.3}$$

The crucial point is that in a numerical calculation, the two values of the amplitude are computed with an entirely different floating point arithmetic. This leads to two independent calculations although algorithmically, exactly the same happens. The only requirement is that the scaling parameter is not an integer power of two since this would only lead to a shift of the exponent in the binary representation of the floating point number while the arithmetic operations in the mantissa would be exactly the same. A "non-binary" scaling changes both the mantissa and the exponent of the floating point number leading to a different floating point arithmetic. This means that those digits of the final result that are affected by rounding errors or numerical instabilities will in general change. To make a quantitative statement on the reliability of the scaling test, we compare for 6-gluon amplitudes with different helicity configuration the accuracy obtained from the scaling test δ_s with the one obtained from the comparison with analytical formulae δ_a taken from Refs. [140, 144]. In Fig. 3.1, the deviation of the two uncertainties is plotted with help of the formula

$$\text{deviation} = \frac{\delta_s}{\delta_a} - 1 \tag{3.4}$$

for 50000 phase space points with the same kinematic cuts as given in Ref. [72]. Zero deviation is thus found if the two estimates lead exactly to the same result, i.e. if they are 100% correlated. As expected, this is not the case for all events, however, the central peak around zero and the small width of the distribution demonstrate that the scaling test can indeed be used as a measure of the accuracy if no analytic results are available. Analysing Fig. 3.1 in more detail, one can see that the centre of the distribution is not exactly located at zero but slightly shifted to the right. In principle this is a hint that the scaling test has a tendency to return a higher accuracy than the analytic formulae. First of all, the effect is tiny and will thus not question the validity of the scaling test as a whole. One source for this shift could be the incorrect assumption that the error

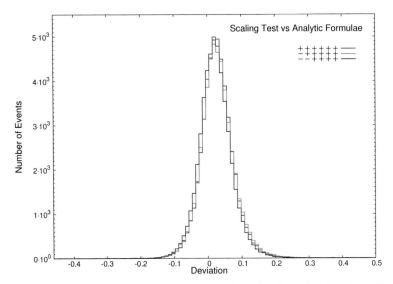

Figure 3.1.: Validation of the scaling test. The distribution shows the relative deviation of two different uncertainty estimates δ_a and δ_s as defined in Eq. (3.4) for various helicity configurations. δ_a is determined with analytic formulae taken from Refs. [140, 144], and δ_s with the scaling test defined in Eq. 3.3.

estimate with analytical formulae is always correct. In addition, it is possible that one estimates a higher accuracy than one actually has if one digit agrees accidentally. This can happen both for δ_a and δ_s. In combination with the bias that the analytic formulae give at any time a correct answer, this could lead to the observed shift in one direction. From all ways to access the numerical uncertainty that we have checked, the scaling test was the only method to get reliable accuracy estimates. Although it doubles the runtime, this time is well invested to have the numerics under control. We stress once again that the scaling test is exclusively based on scaling properties of the amplitude and, therefore, independent of the underlying algorithm.

In Fig. 3.2 the accuracy distribution for a 6-gluon primitive amplitude with MHV helicity configuration $-- + + + +$ over one million phase space points with the same kinematic cuts as given in Ref. [72] is shown. The finite part has been investigated with the scaling test and the poles have been determined by comparison with the analytic structure. We observe that the accuracy of the $\frac{1}{\epsilon}$-pole follows roughly the same pattern as the finite part. Although this looks like a good correlation on the level of the whole distribution, we could confirm at individual phase space points that this is not necessarily true event by event. Note that this is a logarithmic plot, the shape of the distribution might thus be misleading: In fact, less than one permille of the phase space points have an accuracy above -3 (worse than permille accuracy) which is suffi-

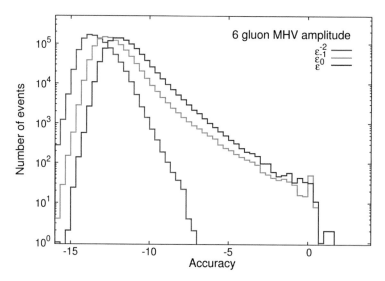

Figure 3.2.: Accuracy distribution of a 6-gluon primitive amplitude with $(- - + + + +)$ MHV helicity configuration. The finite part is checked with the scaling test.

cient for most (if not all) practical applications at the LHC. The "steep fall-off" around zero accuracy is related to the definition of the accuracy in Eq. 3.2: As soon as no digit agrees, it does not make sense any more to ask how "wrong" the result actually is, thus the bins usually vanish close above zero accuracy.

To access the dependence on the multiplicity in a quantitive way, we give in Tab. 3.1 the fraction of events with an accuracy above -3 determined via the scaling test as a function of the number of external gluons, again using the MHV configuration and the phase space cuts from Ref. [72]. It is interesting to note that in the 4-point case there is not a single bad event among one million phase space points. Apart from

n gluons	bad points [%]
4	—
5	0.03
6	0.06
8	0.2
10	0.8
12	3.

Table 3.1.: Fraction of events with an accuracy above -3 for MHV amplitudes as a function of the multiplicity.

the much simpler diagrammatic and kinematic structure of the amplitude, this is most likely related to the absence of pentagon contributions, one of the main sources of numerical instabilities. From Tab. 3.1 one can read off that increasing the multiplicity by one additional gluon roughly doubles the number of bad points. The same could be observed also by requireing $\delta = -4$, though the fraction of bad points for fixed multiplicity is doubled with respect to $\delta = -3$. To get a feeling up to which multiplicity one

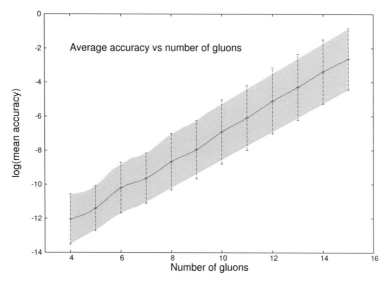

Figure 3.3.: Average accuracy of an MHV amplitude as a function of the multiplicity. The error bars respectively the blue band show the width of the distribution for fixed multiplicity.

can expect to get a reliable answer using mainly built-in double precision, we illustrate in Fig. 3.3 the average accuracy of the finite part of an MHV amplitude determined with the scaling test as a function of the multiplicity for a fixed number of phase space points. We employ the same phase space cuts as in the previous examples. The error bars respectively the blue band show the width of the accuracy distribution computed as the mean square deviation from the average accuracy. To a good approximation, we observe a linearly raising behaviour from around 12 valid digits for four external gluons up to 3-4 valid digits for 14 external gluons. Although the width increases for higher multiplicities as well, we can read off that up to 12-14 gluons, still most of the calculated events have around four significant digits which is enough for most practical applications at the LHC. For higher multiplicities, the fraction of events that require extended precision becomes dominant. Most of the analysis for varying multiplicity is based on the MHV helicity configuration because in addition to the scaling test, we used also the analytic all-n-formula of Ref. [140] for the rational part to compare the numerical results with. Investigating with the scaling test also the NMHV $---+\ldots+$

and the alternating helicity configuration, we found only a weak dependence on the helicity.

The helicity dependence is analysed once again separately using analytic formulae instead of the scaling test. Fig. 3.4 shows a comparison between the numerical results from NGLUON with numerically evaluated analytic formulae for the rational part of 6-gluon primitive amplitudes for different helicity configurations. The plot is over one million phase space points where we have again applied the same cuts on the phase space as in Ref. [72]. The analytic formulae are taken from Refs. [67, 140, 144]. The ra-

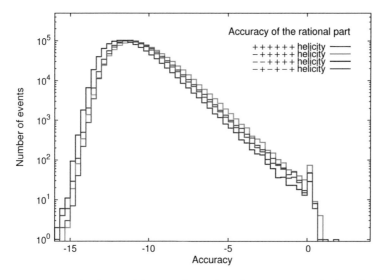

Figure 3.4.: Accuracy distribution of the rational part of a 6-gluon primitive amplitude with different helicity configurations.

tional part is the numerically most challengeing contribution because first, it includes also the less stable pentagons and because second, it involves a much more complicated subtraction procedure than present in the cut-constructible part. We can take it, therefore, as a worst case scenario. The observed distributions look roughly the same for the different helicity configurations. Again, the fraction of events with an accuracy above -3 is less than one permille. The observed excess in the histogram around $\delta = 0$ where no digits agree is again due to the definition of the accuracy. This bin somehow accumulates most of the mismatched points. Fig. 3.5 shows as well a comparison between analytic formulae and results from NGLUON, though at this time the rational part of an eight-point primitive with one external quark line and six gluons with helicity configuration $- + \ldots + $ for different quark–anti-quark separations (the anti-quark always being in the first position). The analytic formula is taken from Ref. [141]. The upper plot shows the mixed quark–gluon loop and the lower plot the case of a closed

quark loop. The fraction of events with an accuracy above -3 is at most at the order of permille, however, the accuracy is now sensitive both to the loop content and the configurations of the external quark pair. For the mixed quark–gluon case, we observe that the accuracy is to a large extent the same for different quark–anti-quark separations. The contrary is the case for the closed quark loop where the number of common digits increases with larger $\bar{q}q$-separation. The reason for this behaviour is that the fermion loop primitive amplitudes have fewer allowed propagators in the parent diagram for increasing $\bar{q}q$-separation. Hence, the number of vertices and integral coefficients is reduced leading to a less involved numerical calculation with better accuracy.

The accuracy distributions involving external quarks look in general better than those with gluons exclusively. This is also expected because first, the maximum tensor rank is lower and, therefore, the reduction is less involved and second, there are simply less diagrams with less complicated interactions that contribute. This is illustrated in Fig. 3.6, showing the accuracy distributions of 10^4 phase space points for the full colour and helicity summed interference term between one-loop and Born amplitude for two basic channels contributing to four-jet production: the pure gluonic channel $gg \to gggg$ which is the most complicated process and the channel with three different flavour quark lines $d\bar{d} \to \bar{u}u\bar{s}s$ being the simplest one. A cut of $|p_i \cdot p_j| > s \times 10^{-2}$ is applied where $s = 7$ TeV being the centre of mass energy. This cut corresponds to the JADE jet algorithm [145, 146]. The distributions show the deviation of the numerically computed epsilon poles from the analytical structure and the accuracy of the finite part determined with the scaling test. The threshold for reevaluation of the finite part in quadruple precision is set to $\delta = -4$ which concerns all bins in the grey shaded area. For the pure gluonic channel this is around 1% of all phase space points. The reprocessed points drawn in the figure with thin lines are all in the allowed region with enough significant digits. In the quark channel, all distributions are much narrower and no event requires the usage of extended precision. The behaviour of all other channels contributing to four-jet production lies between these two extremes.

All the tests that we have presented so far prove that the implementation per se is correct and the underlying numerical accuracy is well under control for phenomenological applications. Beyond that, we made additional cross-checks against known processes at individual phase space points, either with selected analytical formulae from literature, numerically stated benchmark points or publicly available reference programs. At the level of primitive amplitudes, we could reproduce the exact numerical values of the different 4-gluon, 5-gluon and 6-gluon helicity amplitudes from Ref. [87]. In Ref. [70] numerical values of gluon primitive amplitudes with multiplicities $n = \{6, 7, 8, 9, 10, 15, 20\}$ at given phase space points are stated for different helicity configurations. Computing in double precision, we find good agreement for all the low multiplicity configurations up to $n = 10$. For $n = \{15, 20\}$ the agreement is getting worse. Most likely, the mismatch arises because the benchmark points in Ref. [70] have been computed with quadruple precision, however, the phase space points are given only in double precision with 16 valid decimal digits. Since on-shellness and four-momentum conservation is in this case only satisfied for at most 16 decimal digits,

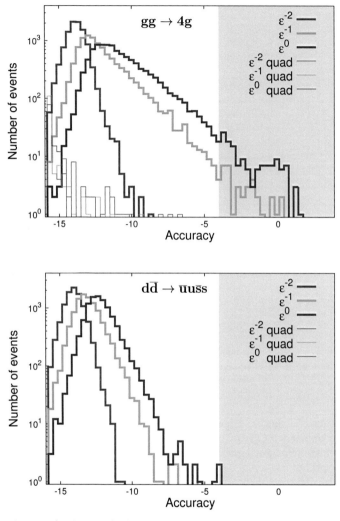

Figure 3.6.: Accuracy distributions of full colour and helicity summed interference terms of virtual and Born amplitude for channels contributing to four-jet production. The most complicated pure gluonic process requires around 1% of the points to be reprocessed in quadruple precision (shaded area) while the much simpler process with three different flavour quark lines does not need any extended precision at all.

	NJET	BLACKHAT
	$gg \to gggg$	
Born	$+4.976935 7371 7948 38 \times 10^8$	$+4.976935736 \times 10^8$
$1/\epsilon^2$	$-1.7999999999 99108 \times 10^1$	-1.800000000×10^1
$1/\epsilon$	$-6.5144862057 710100 \times 10^1$	-6.514486205×10^1
finite	$-3.2130366101 334992 \times 10^1$	-3.213036625×10^1
	$q\bar{q} \to gggg$	
Born	$+2.1622011190 045194 \times 10^5$	$+2.162201118 \times 10^5$
$1/\epsilon^2$	$-1.4666666666 663076 \times 10^1$	-1.466666667×10^1
$1/\epsilon$	$-5.8264471151 950865 \times 10^1$	-5.826447114×10^1
finite	$-4.3957884552 089730 \times 10^1$	-4.395788455×10^1
	$q_1\bar{q}_1 \to q_2\bar{q}_2 gg$	
Born	$+1.3745823177 248822 \times 10^4$	$+1.374582317 \times 10^4$
$1/\epsilon^2$	$-1.1333333333 333261 \times 10^1$	-1.133333333×10^1
$1/\epsilon$	$-4.8019618061 344993 \times 10^1$	-4.801961805×10^1
finite	$-3.7348157184 728159 \times 10^1$	-3.734815718×10^1
	$q_1\bar{q}_1 \to q_2\bar{q}_2 q_3\bar{q}_3$	
Born	$+3.6761219414 819656 \times 10^1$	$+3.676121941 \times 10^1$
$1/\epsilon^2$	$-8.0000000000 001386 \times 10^0$	-8.000000000×10^0
$1/\epsilon$	$-2.6337813992 821804 \times 10^1$	-2.633781399×10^1
finite	$-6.7242846898 320696 \times 10^{-1}$	$-6.724284689 \times 10^{-1}$

Table 3.2.: Numerical comparison between NJET and the $2 \to 4$ amplitudes from BLACKHAT given in Ref. [98].

switching to quadruple precision improves the agreement only moderately. Primitive amplitudes with a massless closed quark loop and exclusively external gluons have been checked against all benchmark points given in Ref. [147] (the phase space points and helicity configurations are the same as in Ref. [70], though stated only for multiplicities $n = \{6, 7, 8, 9, 10\}$). We find numerical agreement. In the case of primitives with one external quark–anti-quark pair, we checked the implementation against the analytic formulae of the five point process $\bar{q}q+3g$ given in Ref. [74]. For all independent helicity configurations and quark–anti-quark separations we find agreement.

At the level of full colour and helicity summed amplitudes, we found agreement with the analytic results for two jet production in Ref. [148]. In addition, HELAC-1LOOP [83] was used to generate numerical cross checks for the colour summed interferences of almost all processes which contribute to three- and four-jet production. The only missing contribution was the six-gluon channel which was not accessible with HELAC-1LOOP. The fermion loop contributions of the six-point processes with external quark lines could also successfully be compared with results generated from GOSAM [89]. Further, we recovered all numerical values at a single phase-space point of the channels contributing to four-jet production provided by BLACKHAT [98]. The explicit comparison is summarised in Tab. 3.2. Except in the pure gluonic channel, we recover the ten decimal digits from [98] using solely double precision. Investigating the pure gluonic channel in more detail, we can estimate the numerical uncertainty of the finite par with our internal scaling test

$$A_{6g}^{(1),\text{finite}} = -3.2130366101334992 \times 10^1 \pm 3 \times 10^{-8},$$

indicating that all significant digits obtained from NJET using double precision agree with the results from BLACKHAT. Switching to quadruple precision, one obtains

$$A_{6g}^{(1),\text{finite}} = -3.2130366250275191 \times 10^1 \pm 2 \times 10^{-13}$$

hence we recover indeed all digits from [98].

All cross-checks for pure gluon primitive amplitudes are accessible via the NGLUON library with the demo program NGluon-demo. The checks for primitive amplitudes with external fermions are accessible from the NJET library with the demo program NParton-qqchecks and all checks at the level of colour summed interferences via the sample program NJet-demo of the NJET library.

We employed the NJET library to compute cross sections and differential distributions for three-jet and four-jet production at 7 and 8 TeV centre of mass energy presented in Ref. [91]. The 7 TeV case has been previously studied by the BLACKHAT collaboration [98]. The fact that we find excellent agreement with their phenomenological results is an additional strong check of the implementation.

3.2. Runtime performance

Discussing the runtime, we are interested in two basic quantities: First, the scaling behaviour of the algorithm as a function of the multiplicity, and, second, the absolute timings for the matrix element evaluation of those processes which are relevant for phenomenology. For this purpose, we discuss first the scaling behaviour of primitive amplitudes and turn afterwards to the runtime of full colour and helicity summed interferences between Born and one-loop amplitude.

The asymptotic behaviour of the implemented algorithm for evaluating primitive amplitudes depends on the number of tree-amplitudes that need to be computed to reconstruct the integrand and on the scaling behaviour of the tree-amplitudes that enter unitarity cuts. The exact number of trees depends also on the flavour of the particles circulating in the loop (for example if the parent diagram exhibits forbidden propagators in the presence of external quarks, there are less integral coefficients to compute). However, to understand the overall asymptotic behaviour, it is sufficient to investigate the scaling for external gluons exclusively. We will explain below that the inclusion of external quarks leads on average to the same asymptotic behaviour. Treating the cut-constructible part and the rational part separately, the total number of trees is given by

$$\mathcal{N}_{\text{tot}} = \mathcal{N}_{\text{cc}} + \mathcal{N}_{\text{rat}} \tag{3.5}$$

$$\mathcal{N}_{\text{cc}} = S_{\text{cc}} \left(4 f_4^{\text{cc}} \binom{n}{4} + 3 f_3^{\text{cc}} \binom{n}{3} + 2 f_2^{\text{cc}} \left[\binom{n}{2} - n \right] \right) \tag{3.6}$$

$$\mathcal{N}_{\text{rat}} = S_{\text{rat}} \left(5 f_5^{\text{rat}} \binom{n}{5} + 4 f_4^{\text{rat}} \binom{n}{4} + 3 f_3^{\text{rat}} \binom{n}{3} + 2 f_2^{\text{rat}} \left[\binom{n}{2} - n \right] \right) \tag{3.7}$$

S_w with $w \in \{\text{cc}, \text{rat}\}$ are the number of possible helicity configurations per subamplitude within the loop. In the pure gluonic case, we have $S_{\text{cc}} = (D_s - 2)^2 = 4$ and $S_{\text{rat}} = 1$. In order to analyse the asymptotic behaviour, we may divide S_{cc} by a factor of 2 because with off-shell recursion, we compute amplitudes which differ only in the helicity of the last leg basically at the same time, c.f. the discussion in appendix B.1. f_R^w denotes the total number of Fourier modes for a topology R to disentangle the integrand. In NGLUON, we have for the cut-constructible part $f_4^{\text{cc}} = 2$, $f_3^{\text{cc}} = 7$ (respectively $f_3^{\text{cc}} = 8$, c.f. the discussion in section 2.5), $f_2^{\text{cc}} = 15$ and for the rational part $f_5^{\text{rat}} = 1$, $f_4^{\text{rat}} = 6$, $f_3^{\text{rat}} = 21$ and $f_2^{\text{rat}} = 30$. The integer constants in front of f_R^w are the number of amplitudes per cut. Finally the binomial coefficients denote the number of possible on-shell settings within a distinct topology R. In the bubble case, we took into account that massless on-shell bubbles vanish. The explicit dependence on the multiplicity n enters via the binomial coefficients, hence, for large n, the pentagon contribution will dominate the algorithm and scale as $O(n^5)$. In appendix B.1 it is shown that the asymptotic behaviour of tree amplitudes for unitarity cuts computed via off-shell recursion scales on average as $O(n^3)$ for the mixed quark–gluon case and as $O(n^2)$ for the closed quark loop case. The theoretical prediction for the asymptotic scaling is thus $O(n^8)$ for the mixed quark–gluon case and $O(n^7)$ for the closed quark loop case. Note that our implementation of the unitarity trees leads to an improved scaling with respect to the $O(n^9)$ behaviour stated in Ref. [87] for the pure gluonic case. The authors of Ref. [87] could show that within their framework the asymptotic regime is reached already for multiplicities around $n = 20$. In Ref. [77], we demonstrated that fitting the runtime for up to 20 gluons leads to a polynomial behaviour of $O(n^6)$, hence the asymptotic regime is not yet reached. In order to get a feeling what "large n" means in practice, we expand the above equations for the given numerical constants

$$\mathcal{N}_{\text{tot}} = \frac{1}{24} n^5 + \frac{5}{4} n^4 + \frac{215}{24} n^3 + \frac{95}{4} n^2 - 154 \, n. \tag{3.8}$$

The coefficient in front of the n^5 term is thirty times smaller than the one in front of the n^4 term, i.e. the two terms are equal for $n = 30$ which gives a first explanation why the pentagons are to a large extent suppressed. In addition, one should also remember that the subamplitudes for the rational part are computed with scalar–gluon vertices which, compared with the pure gluonic ones, involve three times less terms (c.f. Eqs. (2.93), (2.94) and Eqs. (1.18), (1.19)) a fact which pushes the asymptotic regime in practice

$\bar{q}q$-separation	0	1	2	3	4	5	6
mixed quark–gluon loop [ms]	46.6	48.6	51.7	55.7	60.4	65.8	71.9
closed quark loop [ms]	50.6	28.5	14.5	6.2	1.8	—	—

Table 3.3.: Runtime measurement of 8-point primitives with one external quark line, fixed helicity and all possible quark–anti-quark separations. The time is given in milliseconds including the scaling test. The data were taken on an Intel(R) Core(TM)2 Duo CPU E8400 @ 3.00GHz processor. The last two amplitudes in the closed quark loop case are skipped since they involve massless on-shell bubbles and massless tadpoles.

even further above. Empirically, we found that even for $n = 60$ the asymptotic regime is not yet reached, although the fitted exponent is continuously increasing towards $n = 8$. Switching all contributions except the pentagons off, one observes indeed an $O(n^8)$ behaviour already for multiplicities around $n = 20$. Similar results are found for solely external gluons with a closed quark loop, where the predicted $O(n^7)$ behaviour is still far beyond $n = 20$.[2]

For primitives with external quarks, the total number of subamplitudes changes because first, S_{rat} in Eq. (3.7) depends now on the flavour structure of the individual subamplitudes and is in general not factored out and because second, for certain quark configurations there are simply less possibilities to set propagators on-shell. Both effects are illustrated in Tab. (3.3) showing the runtime of eight-point primitives with one external quark pair for different quark–anti-quark separations, fixed helicity $- + \ldots +$ and the amplitude cache described in section 2.7 switched off. The timings (given here and for the remainder of this section) are taken with the scaling test turned on which doubles the runtime, c.f. the discussion in section 3.1. For larger $\bar{q}q$-separation, the runtime increases in the mixed quark–gluon case. This is mainly because in the rational part the number of scalar propagators decreases while the number of massive quark propagators with additional spin sums increases leading to more subamplitudes. In the closed quark loop case, we observe the opposite pattern: The runtime decreases for larger $\bar{q}q$-separation which is a direct consequence of the fewer allowed propagators in the parent diagram. The last two separations are not computed since they would involve massless on-shell bubbles and massless tadpoles. Note that only for the neighbouring quark–anti-quark pair (the leading colour contribution), the evaluation of the closed quark loop is slower than the mixed quark–gluon case. While different flavours in the parent diagram of a primitive amplitude modifiy only the numerical prefactors in Eq. (3.8), the presence of a too large number of forbidden propagators can also dampen the polynomial scaling. For example, a closed fermion loop with a single quark line and maximal $\bar{q}q$-separation involves at most triangle topologies which implies that the total number of trees scales as $O(n^3)$ instead of $O(n^5)$. Averaging over all possible quark–anti-quark separations, however, leads asymptotically still to the $O(n^8)$

---

[2]In practical calculations, we do not compute the rational part of gluon amplitudes with fermion loop by means of massive quarks. Instead, we exploit the supersymmetric decomposition given in Eqs. (2.90), (2.91), hence the $O(n^7)$ scaling is in this case more of academic nature.

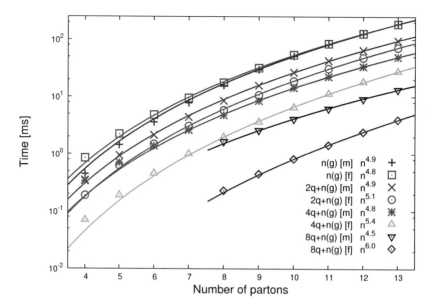

Figure 3.7.: Runtime behaviour of primitive amplitudes with 0, 1, 2 and 4 external quark lines averaged over all possible helicity configurations. Exclusively neighbouring quark pairs for both mixed quark–gluon and closed fermion loop are considered. Both the cache system and the scaling test are switched on. The data are fit to a polynomial of the form $f(n) = a\,n^b$ where the exponent can be read off from the legend. The data were taken on an Intel(R) Core(TM)2 Duo CPU E8400 @ 3.00GHz processor.

and $O(n^7)$ behaviour already discussed for the case of external gluons exclusively.

Fig. 3.7 illustrates the mean runtime per primitve amplitude averaged over all possible helicity assignments, including those configurations which are set equal to zero due to helicity violation along the quark line. The cache system for subamplitudes is switched on for the measurement. The maximum multiplicity $n = 13$ is chosen because this is the highest allowed number of partons for the amplitude cache, c.f. the discussion in section 2.7. The plot shows configurations with 0, 1, 2 and 4 external quark lines, at any time the mixed quark–gluon and the closed quark loop case. Only neighbouring quarkpairs are considered. We observe the expected pattern that the more external quark lines enter, the faster the evaluation is. The reason is again the reduction in complexity due to fewer interaction vertices and, in the case of a closed quark loop, fewer integral coefficients to compute due to forbidden propagators in the parent diagram. In contrast to the timings of a single amplitude evaluation in Tab. 3.3 where the closed quark loop with neighbouring quark configuration is still slower than the mixed quark-gluon case, the helicity average in combination with the amplitude cache

Process	T_d^c	T_d	T_q^c	T_q	T_q/T_d	T_d/T_d^c	T_q/T_q^c	X_d/T_d^c
4g	0.04	0.06	0.35	0.65	10.8	1.5	1.8	0.20
2q2g	0.04	0.08	0.34	0.80	10.3	1.9	2.4	0.24
4q	0.01	0.02	0.13	0.19	10.7	1.2	1.5	0.20
4q–2	0.03	0.04	0.26	0.42	10.5	1.2	1.6	0.24
5g	0.34	0.54	2.74	5.72	10.6	1.6	2.1	0.27
2q3g	0.49	0.95	4.00	9.82	10.4	1.9	2.5	0.23
4q1g	0.16	0.23	1.39	2.32	10.0	1.4	1.7	0.14
4q1g–2	0.32	0.45	2.68	4.63	10.2	1.4	1.7	0.18
6g	9.0	17.8	70.4	189.0	10.6	2.0	2.7	0.28
2q4g	10.5	23.3	81.6	236.7	10.2	2.2	2.9	0.25
4q2g	3.1	5.6	25.3	56.3	10.1	1.8	2.2	0.19
4q2g–2	6.0	11.2	49.1	112.8	10.0	1.9	2.3	0.20
6q	0.5	0.8	4.8	8.1	10.0	1.4	1.7	0.16
6q–2	1.1	1.6	9.3	16.1	9.9	1.5	1.7	0.16
6q–3	3.2	4.8	26.8	48.3	10.0	1.5	1.8	0.18
7g	245	554	1868	5756	10.4	2.3	3.1	0.29
2q5g	267	668	2036	6592	9.9	2.5	3.2	0.25
4q3g	66	145	534	1406	9.7	2.2	2.6	0.19
4q3g–2	133	290	1055	2812	9.7	2.2	2.7	0.20
6q1g	12	21	98	205	9.6	1.8	2.1	0.15
6q1g–2	24	43	1923	411	9.6	1.8	2.1	0.16
6q1g–3	69	129	557	1233	9.6	1.9	2.2	0.18

Table 3.4.: Runtime measurements for full colour and helicity summed interferences between Born and virtual corrections including the scaling test. For multiple fermion channels, an additional tag in the process list denotes the number of equal quark flavours. The first four colums show the timings in seconds for double precision with cache T_d^c, double precision without cache T_d, quadruple precision with cache T^c and quadruple precision without cache T_q. The remaining colums show different ratios of the timings, X_d is the estimated time spent within the cache according to Eq. (3.10). The measurements have been performed on a computer with an Intel(R) Core(TM) i5-2320 CPU @ 3.00GHz processor and 8 GB RAM.

leads to a behaviour that the evaluation of the closed quark loop is now always faster in the presence of external quark pairs. The impact of the cache for the closed quark loop is indeed greater, because more subamplitudes per cut can be reused in the more cumbersome internal spin sums of the rational part. The measured runtime is fitted to a polynomial of the form $f(n) = a\,n^b$, the exponent can be read off from the legend. However, due to the limited number of allowed partons with the cache system turned on, these numbers serve more as a trend than as the true asymptotic behaviour.

Tab. 3.4 shows runtime measurements for full colour and helicity summed interferences between Born and virtual corrections including the scaling test. The second and the fourth column show the timings in double precision T_d respectively in quadruple precision T_q with the cache system described in section 2.7 turned *off*. The ratio T_q/T_d in the fifth column is to a good approximation constant, hence the floating point arithmetic of quadruple precision is roughly ten times slower than in double precision. The timings with the cache system turned *on* are given in the first T_d^c and third T_q^c column. From the ratios T_d/T_d^c and T_q/T_q^c in the sixth and seventh column one can read off a speed gain varying between a factor 1.2-2.5 in double precision and 1.5-3.2 in quadruple precision. The speed-up is process dependent, as expected, and increases with higher multiplicity. The latter is mainly because the average multiplicity of a cached subamplitude increases, thus, the evaluation of amplitudes becomes more important. The fact that the speed-up ratio of a single process is 15-35% larger in quadruple precision than in double precision is an indication that the cache system itself has a non-negligible overhead. To estimate the percentage of the time spent within the cache per phase space point in single precision, we split the measured time with the cache turned on T^c in the time spent with floating point arithmetic T^f and the time X for the cache organisation: $T^c = T^f + X$. With the definition $R \equiv T_q/T_d$, one can deduce from the ratio

$$S \equiv \left(\frac{T_q}{T_q^c}\right) \bigg/ \left(\frac{T_d}{T_d^c}\right) = R \frac{T_d^f + X_d}{T_q^f + X_q} \approx R \frac{T_d^f + X_d}{R \cdot T_d^f + X_d} \qquad (3.9)$$

the estimate

$$\frac{X_d}{T_d^c} \approx \frac{S-1}{S - \frac{1}{R}} \qquad (3.10)$$

where we have assumed that X is the same for double and quadruple precision and that $T_q/T_d \approx T_q^f/T_d^f$. The ratio X_d/T_d^c is plotted in the last column of Tab. 3.4. Varying from process to process, around 15-30% of the runtime is spent mainly for searching amplitudes within the cache and rebalancing the binary tree. On a first glance, this might seem much, however, even if we improved this technical overhead, it will not change the order of magnitude of the runtime per phase space point evaluation.

The channels for four-jet production are all in the order of magnitude of a few seconds per phase space point. NJET can be linked via the so called Binoth Les Houches Accord (BLHA) [149] to standard Monte Carlo integration routines. Using NJET in connection with the SHERPA event generator [10] and a threshold for reevaluation in quadruple precision set to less than four valid decimal digits, we found that 10000 phase space points on a Intel(R) Core(TM) i5-2320 CPU @ 3.00GHz need roughly 24 hours evaluation time. We emphasise, however, that this number depends also on the setup of the integration routine and should thus be taken at most as a rough estimate.

4. Four-jet production at next-to-leading order for the Large Hadron Collider

In this chapter, four-jet production at NLO accuracy for the LHC in massless QCD at a centre of mass energy of $\sqrt{s} = 8$ TeV is presented. This process is important for the successful data analysis at the LHC. It serves on the one hand as a background process for new physics searches. On the other hand, it is of particular interest for precision measurements in QCD to constrain the strong coupling constant α_s and the parton distribution functions (PDF). Four-jet production at NLO is very difficult to calculate, mainly, because the hard matrix elements involve exclusively particles that carry colour charge. Besides the very intricate amplitude structures of both virtual and real corrections on its own, the combination of them in a full NLO computation within the Catani-Seymour subtraction scheme [8] requires in addition the evaluation of a huge number of dipoles. The enormous complexity of the calculation is the main reason why only recently, first results of this process at $\sqrt{s} = 7$ TeV were available [98]. We could confirm these results which is per se non-trivial taking the difficulty of the process into account [91]. In addition, we give significant extensions to the results presented earlier. This concerns not only the new results at $\sqrt{s} = 8$ TeV for the total cross section and differential distributions, but we give also new insights in the size of NLO corrections compared to the LO result [91]. We will first describe the outline of a general NLO calculation in section 4.1, the numerical setup is given in section 4.2 and, finally, the results are presented in section 4.3.

4.1. General outline of the computation

The whole computation is performed in the five flavour scheme where all quark masses are set equal to zero. In particular, the bottom quark is included in the initial state. Through the matching of the coupling constants between the five flavour and the six flavour theories, effects due to the top quark mass are retained. The expansion of the n-jet differential cross section in the coupling α_s reads:

$$d\sigma_n = d\sigma_n^{\text{LO}} + d\delta\sigma_n^{\text{NLO}} + \mathcal{O}(\alpha_s^{n+2}) \tag{4.1}$$

where $d\sigma_n^{\text{LO}} \sim \alpha_s^n$ denotes the LO differential cross section and $d\delta\sigma_n^{\text{NLO}} \sim \alpha_s^{n+1}$ is the NLO correction to the differential cross section. The leading order differential cross section is given by

$$d\sigma_n^{\text{LO}} = \sum_{\substack{i,j \\ \in \{q,\bar{q},g\}}} dx_i dx_j \Phi_{j/H_2}(x_j, \mu_f) \Phi_{i/H_1}(x_i, \mu_f) d\sigma_n^{\text{B}} \big(i(x_i P_1) + j(x_j P_2)\big) \to n \text{ part.}). \quad (4.2)$$

P_1 and P_2 are the momenta of the incoming hadrons which we assume to be massless. x_i are the momentum fractions of the initial state partons of flavour i present in the hadrons H_1 and H_2. The total incoming momentum of the initial state partons is thus $P = x_i P_1 + x_j P_2$ leading to the partonic centre of mass energy $\hat{s} = 2x_1 x_2 (P_1 \cdot P_2) = x_1 x_2 s$. Φ_{i/H_k} denotes the parton distribution functions (PDF) which describe qualitatively the probability of finding a parton of flavour i with momentum fraction between x_i and $x_i + dx_i$ within the hadron H_k. μ_f is the factorisation scale. The differential partonic cross section $d\sigma_n^{\text{B}}$ describes the reaction $(i+j \to n \text{ jets})$ in Born approximation with help of the matrix element $\mathcal{M}_n(ij \to n \text{ part.})$ and a suitable jet algorithm $\Theta_{n\text{-jet}}$

$$d\sigma_n^{\text{B}} = \frac{1}{2\hat{s}} \prod_{\ell=1}^{n} \frac{d^3 k_\ell}{(2\pi)^3 2E_\ell} (2\pi)^4 \delta\Big(P - \sum_{m=1}^{n} k_m\Big) |\mathcal{M}_n(ij \to n \text{ part.})|^2 \Theta_{n\text{-jet}}(k_1, \dots, k_n)$$

$$(4.3)$$

where k_i are the four-momenta of the outgoing partons. The jet algorithm $\Theta_{n\text{-jet}}$ depends on the final state partons and on parameters which define the geometric properties of the jets. The functional value of $\Theta_{n\text{-jet}}$ is equal to one if the momentum configuration corresponds to a valid n-jet event and zero otherwise.

The NLO correction to the differential cross section includes both the virtual corrections $d\sigma_n^{\text{V}}$ and the real corrections $d\sigma_{n+1}^{\text{R}}$ with one additional emitted parton in the final state. While $d\sigma_n^{\text{V}}$ and $d\sigma_n^{\text{B}}$ have the same kinematics, $d\sigma_n^{\text{R}}$ is computed in an extended phase space with different kinematics. Both $d\sigma_n^{\text{V}}$ and $d\sigma_{n+1}^{\text{R}}$ have soft and collinear divergences. They cancel after integrating over the phasespace(s) and factorising the initial state singularities. A method to achieve this technically is the framework of the Catani-Seymour dipole subtraction scheme [8]. The idea is to add and subtract local counterterms which mimic pointwise the singular behaviour of $d\sigma_{n+1}^{\text{R}}$ on the one hand, and which are, on the other hand, chosen such that the additional one-particle phase space leading to the singularity can be integrated analytically to cancel exactly the remaining divergences from the virtual corrections. The NLO correction to the total cross section can thus schematically be written as

$$\delta\sigma^{\text{NLO}} = \int_n (d\sigma_n^{\text{V}} + \int_1 d\sigma_{n+1}^{\text{S}}) + \int_n d\sigma_n^{\text{Fac.}} + \int_{n+1} (d\sigma_{n+1}^{\text{R}} - d\sigma_{n+1}^{\text{S}}), \quad (4.4)$$

where $d\sigma_{n+1}^{\text{S}}$ denotes the local counterterm introduced in the Catani-Seymour subtraction scheme and $d\sigma_n^{\text{Fac.}}$ is due to the factorisation of the initial state singularities. This suggests to split the NLO correction into three different contributions:

$$d\delta\sigma_n^{\mathrm{NLO}} = d\bar{\sigma}_n^{\mathrm{V}} + d\bar{\sigma}_n^{\mathrm{I}} + d\sigma_{n+1}^{\mathrm{RS}}, \tag{4.5}$$

$d\bar{\sigma}_n^{\mathrm{V}}$ is the finite part of the virtual corrections, $d\sigma_{n+1}^{\mathrm{RS}}$ are the real corrections with the local counterterms subtracted and $d\bar{\sigma}_n^{\mathrm{I}}$ contains the finite part of the integrated subtraction terms together with finite remnants from factorisation. All contributions are finite and can be integrated numerically in four dimensions with a Monte Carlo program.

We use the SHERPA Monte Carlo event generator [10] as integration routine for the LO and the NLO cross sections. The Born matrix elements and the tree-level input for $d\bar{\sigma}_n^{\mathrm{I}}$ and $d\sigma_{n+1}^{\mathrm{RS}}$ are evaluated with AMEGIC++ [2] a matrix element generator which is included in SHERPA. Cross checks for the matrix elements have been made with COMIX [3]. In SHERPA, the Catani-Seymour dipole formalism is already implemented. With the help of the Binoth Les Houches Accord interface [149], the virtual corrections obtained from NJET can directly be linked to SHERPA.

4.2. Numerical setup

Throughout the calculation we used the MSTW2008 PDF set [150]. The PDFs are available in different orders — leading order, next-to-leading order and next-to-next-to-leading order — depending on the order used for the evolution and on the order of the hard matrix elements that had been used to fit the PDFs to data. As a consequence, the value of $\alpha_s(\mu_R)$ for fixed μ_R differs depending on the order of the PDF set. For the MSTW2008 PDFs the LO value is $\alpha_s^{\mathrm{MSTW2008\text{-}LO}}(\mu_R = m_Z) = 0.13939$ while at NLO it is $\alpha_s^{\mathrm{MSTW2008\text{-}NLO}}(\mu_R = m_Z) = 0.12018$. The mass of the Z-boson is $m_Z = 91.1876 \pm 0.0021$ [151]. In a LO PDF fit, sizeable corrections of the hard scattering matrix element from higher orders are not taken into account. The fit can still partially compensate for the missing higher orders, though, at the price of a different value for α_s^{LO}. While the deviation from the world average of α_s stated by the particle data group $\alpha_s(\mu_R = m_Z) = 0.1184 \pm 0.0007$ [151] is in the NLO case 1.5%, it is around 10 times larger in the LO case. We use the default setup to evaluate the LO cross section with LO PDFs and α_s^{LO}, while for the NLO cross sections we employ NLO PDFs with α_s^{NLO}. When we discuss the size of the corrections we will come back to this point.

The PDFs and α_s depend on the unphysical factorisation and renormalisation scales μ_f and μ_R which are set in the following to $\mu_f = \mu_R = \mu$. The behaviour of the perturbation series is sensitive to the scale choice. For a fixed value of μ, for example, in extreme regions of phase space where a momentum scale Q differs several orders of magnitude from μ, the occurrence of large logarithms $\sim \log(\mu/Q)$ could spoil the convergence. For this reason, we employ a phase space dependent scale based on the sum of the transverse momenta of the final state partons

$$\hat{H}_T = \sum_{i=1}^{N_{\text{parton}}} p_{T,i}^{\text{parton}}.$$

(4.6)

In particular, we set $\mu = \hat{H}_T/2 \equiv \mu_c$, the central scale. To estimate the effect of uncomputed higher orders in the perturbative expansion, we perform a scale variation multiplying respectively dividing the central scale by a factor of two, i.e. the observables are also evaluated at the upper scale $\mu = \hat{H}_T \equiv \mu_+$ and at the lower scale $\mu = \hat{H}_T/4 \equiv \mu_-$.

For the jet defining function $\Theta_{n\text{-jet}}$ in Eq. (4.3), we use the anti-k_t jet algorithm [152] as implemented in FASTJET [153]. The jet-radius parameter R is set to

$$R = 0.4$$

(4.7)

following the value adopted by the ATLAS collaboration. Events were generated using identical cuts to that of the multi-jet measurements from ATLAS [154]. In particular the transverse momentum, p_T, of the first jet is required to be larger than 80 GeV with subsequent jets required to have at least $p_T > 60$ GeV. Rapidity cuts of $|\eta| < 2.8$ were also taken.

4.3. Results

For the total cross section of four-jet production at 8 TeV we find

$$\sigma_{4\text{-jet}}^{\text{LO}} = 14.36(0.01)_{8.76(-)}^{24.74(+)} \text{ nb},$$

(4.8)

$$\sigma_{4\text{-jet}}^{\text{NLO}} = 8.15(0.09)_{4.91(-)}^{7.91(+)} \text{ nb}.$$

(4.9)

The numbers in parenthesis give the statistical error of the cross section due to the Monte Carlo integration. While this is less than one permille in the LO case, it is around one percent for the NLO result. The numbers which follow immediately after the equality sign in Eqs. (4.8) and (4.9) are the numerical values for the cross section evaluated at the central scale μ_c, the superscripts denote the results obtained at the upper scale μ_+ while the subscripts state the numbers from the lower scale μ_-. First of all, we observe that the NLO correction $\delta\sigma_{4\text{-jet}}^{\text{NLO}}$ to the total cross section is around -45% of the LO result, a sizeable reduction. Looking at the mere numbers including the scale variation, the scale uncertainty is significantly reduced as well at NLO. At LO, one observes that the central value of the cross section is enclosed between the values of the upper and the lower scale choice in a broad band of around 16 nb. In addition to the relative numerical reduction of the scale uncertainty at NLO, the numerical values for both upper and lower scale choice are below the central value. This is what one can consider as the ideal outcome of an NLO calculation: The inclusion of higher order corrections stabilises the LO result. While the LO cross section exhibits a monotone dependence on the scale choice, the NLO cross section as a function of μ approaches a "plateau region" that falls off moderately in in both directions from a maximum value.

	gg	qg	qq
relative contribution	37%	49%	14%

Table 4.1.: Relative contribution of the different parton channels to $pp \to 4$ jets at LO.

An interesting analysis represents the contribution of the individual partonic channels contributing to four-jet production as shown in Tab. 4.1. In this table, neither quarks and anti-quarks in the initial state nor different quark flavours are distinguished from each other. For example the qq channel includes the contribution from the initial states $q\bar{q}', qq', \bar{q}\bar{q}'$. Although the ratios do not form an observable, they give a feeling how the parton flux from the PDFs combines with the weight from the matrix elements in the partonic cross sections. As can be seen, the dominant channels are those with one gluon and one quark in the initial state, a consequence of the largest parton luminosity. Restoring the flavour information, 31% of the 49% from these channels come from $gu \to u + 3g$ and $gd \to d + 3g$. The largest single process is as expected the pure gluonic one $gg \to 4g$ which contributes 30%.

Fig. 4.1 shows the differential distribution of the transverse momentum p_T for the leading jet which is defined as the jet with the numerically largest p_T value The bold lines in the upper part of the plot are the LO (blue) and NLO (red) results evaluated at the central scale μ_c. The scale variation is included at LO as the blue band and at NLO as the red band. For the NLO result, the statistical error of the Monte Carlo integration is also included. Like in the inclusive total cross section, one observes a significant reduction of the scale uncertainty at NLO compared to the LO results. While at LO, for every bin the up and down variation enclose the central value, at NLO, in almost all bins both the up and down variation is smaller than the central value. The width of the red band describing the upwards variation is in these cases set equal to zero while the width of the downwards variation is defined as $\min(d\sigma^{LO}(\mu_+)/dp_T, d\sigma^{LO}(\mu_-)/dp_T)$, i.e. as the larger deviation from the central value. This shows that the differential distribution stabilises when higher order corrections are included. The black line in the lower part of the plot shows the K-factor, the ratio between NLO and LO cross section at the central scale $(d\sigma^{NLO}(\mu_c)/dp_T)/(d\sigma^{LO}(\mu_c)/dp_T)$. The scale variations are included as well: The read band denotes the NLO scale variations with the upwards and downwards variation as defined above normalised to $d\sigma^{LO}(\mu_c)/dp_T$, while the blue band denotes the corresponding LO scale variation normalised to $d\sigma^{LO}(\mu_c)/dp_T$. The reduction of the NLO result is around 45%. It is remarkable that the K-factor is to a good approximation constant over a large p_T range. This shows that the dynamical scale choice can indeed avoid numerically large logarithmic contributions. For very large and very low p_T values, the spectrum becomes unreliable. In the high p_T range this is due to the low statistics: only a small fraction of events have an extremely high transverse momentum. For low p_T, effects of soft gluons become important which

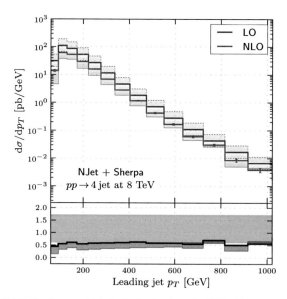

Figure 4.1.: p_T distribution in $pp \to 4$ jets for the leading jet at the LHC with a centre of mass energy of 8 TeV.

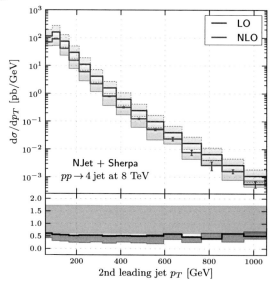

Figure 4.2.: p_T distribution in $pp \to 4$ jets for the second leading jet at the LHC with a centre of mass energy of 8 TeV.

would require going beyond fixed order to take them properly into account. The p_T distribution for the second leading jet is shown in Fig. 4.2. It follows the same pattern as the leading jet, the K-factor is even more stable. Similar results have also been found in the investigation of the third and fourth jet.

In Fig. 4.3, the rapidity distribution of the leading jet is shown, correspondingly the rapidity distribution of the second leading jet in Fig. 4.4. As expected, the distribution is symmetric around the central value $\eta = 0$ and decreases for large $|\eta|$. As far as the behaviour of the corrections is concerned, the qualitative outcome is very similar to the analysis of the transverse momentum: A significant reduction of the scale uncertainty, a reduction of the NLO result around 45-50% and a remarkably constant K-factor. The rapidity distribution of the third and fourth jet follow the same pattern.

When we described the numerical setup for the PDF sets in section 4.2, we mentioned the differing numerical values of α_s depending on the order of the employed PDF set. One may thus ask, to which extent the large negative corrections of the NLO cross section are due to the different value of α_s. To investigate this point in more detail, we evaluate the LO cross section with the corresponding NLO PDF set and the respective α_s^{NLO}. This gives, in addition, also an estimate how the hard matrix element alone changes going from LO to NLO. The results for the transverse momentum and the rapidity distribution of the leading jet are shown in Fig. 4.5. The LO cross section is in this manner divided roughly by a factor of two. If one considers the scale variation as an effect of higher order corrections, it is remarkable that the bands from the LO order scale variation overlap now always with the NLO bands. This was especially for the rapidity distribution in Figs. 4.3 and 4.4 not the case where the default setup for the LO cross section with LO PDFs and α_s^{LO} has been employed. We conclude that the NLO PDF sets with corresponding α_s^{NLO} used for four-jet production at LO order give a phenomenologically better description of what is happening at NLO.

It would be interesting to repeat this analysis also for other processes. Using NLO PDFs for the evaluation of LO cross sections for different process classes could be of general interest for the experimental analysis in cases where the hard matrix elements are not known at NLO or simply take too long to be evaluated. The reasoning behind the larger value of α_s in combination with the LO PDFs is that the PDF fit to the physical data can partially compensate for higher order corrections which are not taken into account via the hard matrix element. The validity for an arbitrary process, however, would require a universal K-factor. Yet, the K-factors of individual processes can be very different, hence, the above mentioned compensation of higher order effects is valid only for a limited number of processes. It is assumed that this difference is enhanced if the order in α_s of the matrix elements that enter the cross sections for the PDF fits is different from the order in α_s for the particular process of interest. Using NLO PDF sets in combination with the LO cross section will thus probably give a better approximation to the physical results of a full NLO computation.

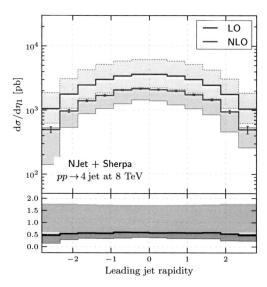

Figure 4.3.: Rapidity distribution in $pp \rightarrow 4$ jets for the leading jet at the LHC with a centre-of-mass energy of 8 TeV.

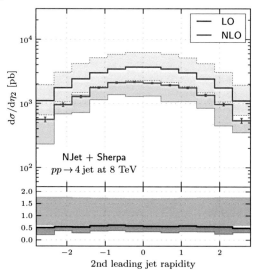

Figure 4.4.: Rapidity distribution in $pp \rightarrow 4$ jets for the second leading jet at the LHC with a centre-of-mass energy of 8 TeV.

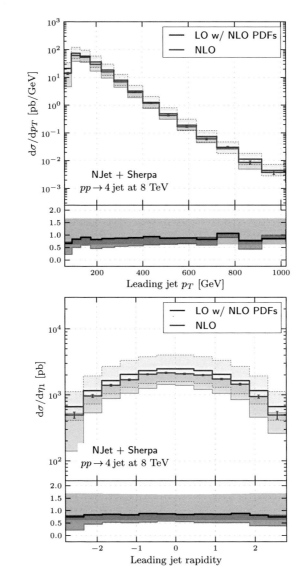

Figure 4.5.: p_T distribution distribution (upper plot) and rapidity distribution (lower plot) of the leading jet in $pp \to 4$ jets using NLO PDFs in the evaluation of the LO cross section, all other numerical setup is left untouched.

Conclusion

In this work, the fully numerical evaluation of virtual corrections for multi-jet production at next-to-leading order in QCD with massless quarks is being presented. The employed method is based on generalised unitarity and integrand reduction techniques and has been implemented in the publicly available computer programs NGLUON and NJET. As a proof of principle for handling a state-of-the-art phenomenological process at the LHC, NJET has been successfully applied to compute four-jet production at next-to-leading order.

In particular, a detailed description for computing primitive amplitudes in massless QCD is being presented. The method combines several aspects of D-dimensional generalised unitarity and integrand reduction for a numerical calculation in four dimensions. This is favourable for computing the rational terms of the amplitude: With dimensional regularisation in the FDH scheme, the degrees of freedom of the dimensional regulator are incorporated in an effective mass shift of the particles circulating in the loop. In addition, all gluons within the loop are replaced by real scalar particles introducing specific scalar–quark vertices. This allows for the reconstruction of the rational part with similar four-dimensional tree-level techniques as it is the case for the cut-constructible part. The presence of massive fermion propagators generates an extended integrand structure with additional spurious terms. We have proven that these terms do not affect the final result as long as the physical quark masses are set equal to zero. The outlined techniques have been successfully automated as a part of the NGLUON and NJET libraries allowing for efficient numerical computation of primitive amplitudes with arbitrary multiplicity, arbitrary helicity configuration and arbitrary number of external quark pairs.

The colour summed interference of the Born amplitude with the full one-loop amplitude is implemented in NJET in terms of primitive amplitudes for all channels contributing to two-jet, three-jet, four-jet and five-jet production. Both at the level of primitive amplitudes and at the level of the colour (and helicity) summed matrix elements, a large number of non-trivial cross checks against known analytic results from literature and against benchmark points at individual phase-space points was performed. This alludes for confidence that the implementation in NGLUON and NJET is correct also for those matrix elements unknown so far.

To estimate the numerical accuracy independently from additional external input, an internal test based on the scaling properties of the amplitude with respect to the external momenta was established. This test has been used to show that the numerics in NJET are well under control: at most 1% of the phase space points of the virtual matrix elements contributing to four-jet production needed to be reprocessed in extended precision (NJET is able to compute in quadruple precision). For the summation over the helicities and permutations of the external particles, a cache system was developed that enables the reuse of tree-level subamplitudes from different unitarity cuts. For four-jet production this resulted in an accelerated runtime by a factor of two. On average, the evaluation of a single-phase space point in double precision with full colour and full helicity is of the order of a few seconds, hence, feasible for a Monte Carlo integration.

Linking NJET with the SHERPA Monte Carlo event generator, we performed a full NLO computation of four-jet production. We found large negative corrections to the LO result to the extent of 45%, both for the total cross section as well as for differential distributions. In addition, the dependence on the unphysical factorisation and renormalisation scale is significantly reduced. Evaluating the LO result with NLO PDFs and the respective α_s^{NLO}, reduces the LO cross-sections roughly by a factor of two. This shows that the large negative corrections arise to a substantial extent from the numerically larger value of α_s^{LO}. As a conclusion, using NLO PDFs and the respective α_s^{NLO} also in the LO cross section gives a better approximation of the full NLO result. It would be of interest to verify this statement also for a larger class of processes since this could give a direction for a better approximation towards the NLO result if the corresponding matrix elements are not available.

The future perspectives are first to exploit NJET for more phenomenology: The full NLO computation of five-jet production at next-to-leading order is currently in progress. We emphasise that the virtual matrix elements of five jet-production (they have been unknown in the literature so far) are all implemented in NJET and are, thus, publicly available now. Second, NJET will be extended to cover more high-multiplicity standard model processes. The inclusion of massive external vector bosons within the present framework is straightforward to achieve. More challenging in this respect is the inclusion of massive external quarks when computing the rational terms. A method to this effect has been suggested in this work and will be subject of a future study.

Appendix

A. Spinor-helicity methods

The basic quantities to construct all external wave functions are two component Weyl spinors denoted by $|p\rangle$ and $|p]$. They are the solution of the two Weyl equations

$$(p \cdot \sigma)|p\rangle = 0 \tag{A.1}$$

$$(p \cdot \bar{\sigma})|p] = 0 \tag{A.2}$$

where p^μ is a light-like four-momentum. The hermitian 2×2 sigma matrices σ^μ and $\bar{\sigma}^\mu$ are defined as

$$\sigma^\mu = (\mathbb{1}_{2\times2}, \vec{\sigma}) \qquad \bar{\sigma}^\mu = (\mathbb{1}_{2\times2}, -\vec{\sigma}) \tag{A.3}$$

with $\vec{\sigma}$ being the Pauli matrices

$$\sigma^1 = \begin{pmatrix} 0 & 1 \\ 1 & 0 \end{pmatrix}, \quad \sigma^2 = \begin{pmatrix} 0 & -i \\ i & 0 \end{pmatrix}, \quad \sigma^3 = \begin{pmatrix} 1 & 0 \\ 0 & -1 \end{pmatrix}. \tag{A.4}$$

The adjoint spinors can be obtained via

$$\langle p| = \left(+ i\sigma_2 |p\rangle \right)^T \tag{A.5}$$

$$[p| = \left(- i\sigma_2 |p] \right)^T \tag{A.6}$$

where T means taking the transpose.[1] We further introduce the light cone coordinates

$$p^+ = p_0 + p_3, \qquad p_\perp = p_1 + ip_2,$$
$$p^- = p_0 - p_3, \qquad \bar{p}_\perp = p_1 - ip_2.$$

---

[1]We avoid in our notation explicitly the usage of dotted and undotted Weyl spinors since it would require the introduction of many additional conventions. The translation into this language reads $\lambda_a = |p\rangle$, $\tilde{\lambda}^{\dot{a}} = |p]$, $\lambda^a = \langle p|$, $\tilde{\lambda}_{\dot{a}} = [p|$. $\pm i\sigma_2$ are the epsilon tensors with upper and lower indices $\epsilon^{12} = -\epsilon^{21} = -\epsilon_{12} = \epsilon_{21} = +1$ which serve as a metric in spinor space to form Lorentz invariant quantities. An excellent review on this theoretically very interesting topic can be found for example in Ref. [155].

In these coordinates, the explicit representation of the Weyl spinors and their adjoints reads

$$|p\rangle = \frac{\sqrt{|p^+|}}{p^+} \begin{pmatrix} p^+ \\ p_\perp \end{pmatrix}, \qquad \langle p| = \frac{\sqrt{|p^+|}}{p^+} \left(p_\perp, \; -p^+ \right),$$

$$|p] = \frac{1}{\sqrt{|p^+|}} \begin{pmatrix} \bar{p}_\perp \\ -p^+ \end{pmatrix}, \qquad [p| = \frac{1}{\sqrt{|p^+|}} \left(p^+, \; \bar{p}_\perp \right). \tag{A.7}$$

They are normalised such that

$$|p\rangle[p| = p \cdot \bar{\sigma}, \qquad |p]\langle p| = p \cdot \sigma. \tag{A.8}$$

Note the asymmetric phase choice between angled and bracket spinors which follows from

$$\frac{\sqrt{|p^+|}}{p^+} = \frac{1}{\sqrt{|p^+|}} e^{-i \arg(p^+)}. \tag{A.9}$$

The phase convention is chosen such that with a momentum flip $p \to -p$ the angled spinors remain unchanged, while the bracket spinors change their sign:

$$|(-p)\rangle = |p\rangle \qquad \langle(-p)| = \langle p| \qquad |(-p)] = -|p] \qquad [(-p)| = -[p|. \tag{A.10}$$

For real momenta and $p^+ \geq 0$, the spinors are related by hermitian conjugation as

$$\langle p| = |p]^\dagger \quad \text{and} \quad [p| = |p\rangle^\dagger. \tag{A.11}$$

For complex momenta, all spinors in Eqs. (A.7) are independent. The Weyl spinors can be used to form Lorentz invariant spinor products

$$\langle p|q\rangle = \sqrt{p^+}\sqrt{p^-} \left(\frac{p^-}{\bar{p}_\perp} - \frac{q^-}{\bar{q}_\perp} \right) \equiv \langle pq\rangle \tag{A.12}$$

$$[p|q] = \frac{1}{\sqrt{p^+}\sqrt{p^-}} \left(p^+\bar{q}_\perp - q^+\bar{p}_\perp \right) \equiv [pq] \tag{A.13}$$

which follows from direct matrix multiplication and the identity $p^+p^- = \bar{p}_\perp p_\perp$. They are anti-symmetric in their arguments: $\langle pq\rangle = -\langle qp\rangle$ and $[pq] = -[qp]$. While also the spinor products may still depend on the normalisation of the spinors, the product of an angled and bracket spinor product does not any more since this is due to Eq. (A.8) exactly the dot product of two Lorentz vectors $\langle pq\rangle[qp] = 2\, p \cdot q$.

For the four component Dirac spinors we use the Weyl representation of the Dirac algebra $\{\gamma^\mu, \gamma^\nu\} = 2g^{\mu\nu}$. The four-dimensional gamma matrices read

$$\gamma^\mu = \begin{pmatrix} 0 & \sigma^\mu \\ \bar{\sigma}^\mu & 0 \end{pmatrix} \qquad \gamma_5 = i\gamma_0\gamma_1\gamma_2\gamma_3 = \begin{pmatrix} -\mathbb{1} & 0 \\ 0 & \mathbb{1} \end{pmatrix} \tag{A.14}$$

where $\{\gamma_5, \gamma^\mu\} = 0$. Those solutions of the massless Dirac equation $\not{p}\psi \equiv p_\mu \gamma^\mu \psi = 0$ which are eigenvectors of γ_5 are called helicity eigenstates. For real momenta, the two different eigenvalues ± 1 define the helicity of the positive energy solutions $\gamma_5 u_\pm = \pm u_\pm$, and the negative eigenvalues define the helicity of the negative energy solutions $\gamma_5 v_\pm = \mp v_\pm$. In particular holds for the adjoint spinors $\bar{u}_\pm \gamma_5 = \mp \bar{u}_\pm$ and $\bar{v}_\pm \gamma_5 = \pm \bar{v}_\pm$. In the Weyl representation where γ_5 is diagonal, the helicity eigenstates can directly be expressed in terms of the two component Weyl spinors. Since we choose all momenta to be outgoing we need in principle only the $\bar{u}_\pm(p)$ spinors for outgoing quarks and the $v_\pm(p)$ spinors for outgoing anti-quarks. The latter explicitly read

$$v_-(p) = \begin{pmatrix} 0 \\ |p\rangle \end{pmatrix} \qquad v_+(p) = \begin{pmatrix} -|p] \\ 0 \end{pmatrix} \tag{A.15}$$

We *define* the massless adjoint four component spinors $\bar{u}_\pm$ as

$$\bar{u}_-(p) = (0, \langle p|) \qquad \bar{u}_+(p) = ([p|, 0) \tag{A.16}$$

We emphasise that we introduced the adjoint spinors by definition, since the usual relation $\bar{u} = u^\dagger \gamma^0$ holds only for real momenta. For complex momenta, the four objects $v_\pm$ and $\bar{u}_\pm$ are all independent and not related by complex conjugation. The phase choice between u and v spinors is such that

$$v_\pm(p) \overset{!}{=} u_\mp(-p) \qquad \text{and} \qquad \bar{u}_\pm(p) \overset{!}{=} \bar{v}_\mp(-p) \tag{A.17}$$

which leads to the following completeness relations:

$$\sum_s u_s(p)\bar{u}_s(p) = \sum_s v_{-s}(-p)\bar{u}_{+s}(+p) = +\not{p} \qquad \begin{array}{c} \overrightarrow{} \\ \overrightarrow{} \pm \mp \overleftarrow{} \end{array} \tag{A.18}$$

$$\sum_s v_s(p)\bar{v}_s(p) = \sum_s v_{-s}(+p)\bar{u}_{+s}(-p) = -\not{p} \qquad \begin{array}{c} \overleftarrow{} \\ \overrightarrow{} \pm \mp \overleftarrow{} \end{array} \tag{A.19}$$

This phase choice is particularly well suited for unitarity computations since it automatically reconstructs the numerators of on-shell fermion propagators correctly with the convention that 1) all tree-level subamplitudes have outgoing momenta and 2) legs of opposite helicity are sewn together. This is diagrammatically shown on the right-hand side of the above equations. The long arrow denotes the parametrisation of the loop momentum flow which can be either parallel or anti-parallel to the fermion flow. The short arrows indicate the momentum flow in the on-shell tree-level subamplitudes.

In order to construct the massive spinors of momentum p with $p^2 = m^2$, we choose massless reference spinors $v(q)$ and $\bar{u}(q)$ with $q^2 = 0$. The spinors that we employ explicitly read

$$v_+(p,q,m) = \frac{-1}{\langle p^\flat q \rangle}(\not{p} - m)\, v_-(q) \qquad v_-(p,q,m) = \frac{-1}{[p^\flat q]}(\not{p} - m)\, v_+(q) \tag{A.20}$$

$$\bar{u}_+(p,q,m) = \frac{1}{\langle qp^{\flat}\rangle}\bar{u}_-(q)(\not{p}+m) \qquad \bar{u}_-(p,q,m) = \frac{1}{[qp^{\flat}]}\bar{u}_+(q)(\not{p}+m) \tag{A.21}$$

The massless projection $p^{\flat}$ of p with respect to q, defined as

$$p^{\flat\mu} = p^{\mu} - \frac{m^2}{2p\cdot q}q^{\mu} \tag{A.22}$$

is constructed such that $p^{\flat}\cdot p^{\flat} = 0$ and $p\cdot q = p^{\flat}\cdot q$. In the massless limit, the massless spinors from Eqs. (A.15) and (A.16) are recovered. The spinors satisfy the completeness relations in full analogy to the massless case:

$$\sum_s v_{-s}(-p,q,m)\bar{u}_{+s}(+p,q,m) = \not{p}+m \tag{A.23}$$

$$\sum_s v_{-s}(+p,q,m)\bar{u}_{+s}(-p,q,m) = -(\not{p}-m) \tag{A.24}$$

Note that the explicit dependence on the reference momentum drops out only after summing over the spins. These spinors have been employed in NGLUON for the computation of tree-level amplitudes for the rational part. In particular, we set $m = \mu$ where μ denotes the additional degrees of freedom from dimensional regularisation.

The suggestion to include also massive external quarks given in section 2.4 requires artificial spinors which depend both on the physical mass m and on the parameter μ which incorporates the degrees of freedom from dimensional regularisation. A particular representation of these objects reads:

$$v_+(p,q,m,\mu) = \frac{-1}{\langle p^{\flat}q\rangle}(\not{p}+i\mu\gamma_5 - m)\,v_-(q) \tag{A.25}$$

$$v_-(p,q,m,\mu) = \frac{-1}{[p^{\flat}q]}(\not{p}+i\mu\gamma_5 - m)\,v_+(q) \tag{A.26}$$

$$\bar{u}_+(p,q,m,\mu) = \frac{1}{\langle qp^{\flat}\rangle}\bar{u}_-(q)(\not{p}+i\mu\gamma_5 + m) \tag{A.27}$$

$$\bar{u}_-(p,q,m,\mu) = \frac{1}{[qp^{\flat}]}\bar{u}_+(q)(\not{p}+i\mu\gamma_5 + m) \tag{A.28}$$

obeying the completeness relations

$$\sum_s v_{-s}(-p,q,m,\mu)\bar{u}_{+s}(+p,q,m,\mu) = \not{p}+i\mu\gamma_5 + m \tag{A.29}$$

$$\sum_s v_{-s}(+p,q,m,\mu)\bar{u}_{+s}(-p,q,m,\mu) = -(\not{p}+i\mu\gamma_5 - m). \tag{A.30}$$

The polarisation vectors for massless spin one gauge bosons of momentum p can be directly constructed in terms of Weyl spinors and a massless reference momentum q

$$\varepsilon_\mu^{(+)}(p,q) = +\frac{\langle q|\bar{\sigma}_\mu|p]}{\sqrt{2}\langle qp\rangle}, \qquad \varepsilon_\mu^{(-)}(p,q) = -\frac{[q|\sigma_\mu|p\rangle}{\sqrt{2}[qp]}. \qquad (A.31)$$

They satisfy the normalisation $\epsilon^{(\pm)} \cdot \epsilon^{(\mp)} = -1$, $\varepsilon^{(\pm)} \cdot \varepsilon^{(\pm)} = 0$, the transversality condition $p^\mu \varepsilon_\mu^{(\pm)}(p,q) = 0$ and the completeness relation

$$\sum_{\lambda=\pm} = \varepsilon_\mu^{(-\lambda)}(p,q)\varepsilon_\nu^{(\lambda)}(p,q) = -g_{\mu\nu} + \frac{p_\mu q_\nu + p_\nu q_\mu}{p \cdot q}. \qquad (A.32)$$

If one evaluates the polarisation vectors with the Weyl spinors given in Eqs. (A.7), then a momentum flip $p \to -p$ reverses at the same time the sign of the polarisation vectors. This is a disadvantage for unitarity calculations where the propagators are reconstructed from polarisation vectors of opposite momentum. In this case, it is advantageous to use Weyl spinors with symmetric phase choices such that the overall phase entirely cancels between numerator and denominator in Eqs. (A.31). A frequently used possibility is to replace $\sqrt{|p^+|}/p^+$ and $1/\sqrt{|p^+|}$ in Eqs. (A.7) with the common normalisation $\exp(-i \arg(p^+)/2)/\sqrt{|p^+|}$, c.f. for example Ref. [156]. The polarisation vectors in NJET are obtained in this way and therefore symmetric under a momentum flip. For all Dirac spinors, we adhere to the explicit formulae of Eqs. (A.7).

B. Algorithmic realisation of colour ordered off-shell recursion

B.1. Constructive implementation of the Berends-Giele recursion

The naive implementation of the Berends-Giele recursion relations in a computer program via a self calling function and a termination condition — a so called top-down approach — is not very efficient since a distinct current appears several times at different stages within the recursion. Thinking of the recursion in terms of a decision tree where one counts every possible branching, it is immediately clear that this would lead to a problem of exponential complexity. We show now that a simple cache system can dampen the asymptotic behaviour to polynomial scaling of $O(n^4)$. The reason why we point out this in principle well established fact [105] is that we can use some elements of the derivation presented here to show that the tree amplitudes needed for unitarity cuts can effectively be computed with a scaling of $O(n^3)$.

To see how the pattern of a cache system emerges, it is best to compute the recursion in a constructive way, a so called bottom-up approach: For the evaluation of an n-point current, one computes first n one-point currents, then $(n-1)$ two-point currents, subsequently $(n-2)$ three-point currents and so on, at any time reusing all known lower point currents. The number of possible currents that may appear in that way is thus $n(n+1)/2$. They can be instructively arranged in an $n \times n$ array of currents where the row number denotes the multiplicity of a current and the column number the index of the first gluon of the current. This leads to an upper triagonal array as in Fig. B.1, the $n(n-1)/2$ lower entries are simply left blank.

$J(1)$	$J(2)$	$\ldots$	$J(n-2)$	$J(n-1)$	$J(n)$
$J(1,2)$	$J(2,3)$	$\ldots$	$J(n-2,n-1)$	$J(n-1,n)$	
$J(1,2,3)$	$J(2,3,4)$	$\ldots$	$J(n-2,n-1,n)$		
$\vdots$	$\vdots$				
$J(1,\ldots,n-1)$	$J(2,\ldots,n)$				
$J(1,\ldots,n)$					

Figure B.1.: Current array which arranges all possible currents that may appear in the computation of the n-point current in a triangular form. The row number denotes the multiplicity, the column number the index of the first gluon in the current.

Assuming the triangular array of off-shell currents to be given, we now want to know the order in n to compute an $(n + 1)$-point current $J(0, \ldots, n)$. In other words, we want to add an additional column of $(n + 1)$ currents to the left-hand side of the array. We further assume for a moment that for an arbitrary $m \in \mathbb{N}$ with $0 \leq m \leq n$ the currents $J(0), J(0, 1), \ldots, J(0, \ldots, m-1)$ were given as well. Ignoring propagators and current indices at the recursion equations (1.17), the $J(0, \ldots, m)$ current is then simply given by

$$J(0, \ldots, m) = \sum_{i=0}^{m-1} V_3 \, J(0, \ldots, i) \, J(i + 1, \ldots, m)$$
$$+ \sum_{i=0}^{m-2} \sum_{j=i+1}^{m-1} V_4 \, J(0, \ldots, i) \, J(i + 1, \ldots, j) \, J(j + 1, \ldots, m)$$

where all currents on the right-hand side are known by assumption. The time for a vertex computation is a constant, or equivalently, its scaling is of $O(1)$. Hence, the order of $J(0, \ldots, m)$ with the above assumptions is simply the number of vertex calls and therefore reads

$$N_{\text{excl}}(0|1, \ldots, m) = \sum_{i=0}^{m-1} 1 + \sum_{i=0}^{m-2} \sum_{j=i+1}^{m-1} 1 = \frac{1}{2} m(m + 1).$$

The subscript 'excl' indicates that all lower point currents are known. The number of vertex calls to fill the column up from $J(0)$ to $J(0, \ldots, m)$ is then simply

$$N_{\text{tot}}(0|1, \ldots, m) = \sum_{i=1}^{m} N_{\text{excl}}(0|1, \ldots, i) = \frac{1}{6} m^3 + \frac{1}{2} m^2 + \frac{1}{3} m \equiv \mathcal{N}_{m \to m+1}. \tag{B.1}$$

Setting $m = n$, we get the number of vertex calls to add $(n + 1)$ off-shell currents to the left of the above triangle array, or, more generally, $\mathcal{N}_{m \to m+1}$ is the number of additional vertex calls going from an arbitrary column of length m to one of length $m + 1$. We can use this relation to compute the total number of vertex calls to fill the $n(n + 1)/2$ elements of the triangular array in Fig. B.1 simply starting from a single one-point current corresponding to $m = 1$:

$$\mathcal{N}_{\text{tot}}[n] = \sum_{m=1}^{n-1} \mathcal{N}_{m \to m+1} = \frac{1}{24} n^4 + \frac{1}{12} n^3 - \frac{1}{24} n^2 - \frac{1}{12} n. \tag{B.2}$$

Hence, filling the current matrix is a process of $O(n^4)$. Since the on-shell contraction of the n-point current with the $(n + 1)$th polarisation vector is independend of the multiplicity, we conclude that the amplitude evaluation with n external legs scales as $O(n^4)$. Although we have analysed the asymptotic behaviour only for the gluonic case, the same behaviour holds also for quark-gluon amplitudes because for sufficiently large n, the recursion will always be dominated by the four-gluon vertex. Yet, to reach the asymptotic regime, higher multiplicities then in the pure gluonic case are required what we could nicely illustrate in Ref. [113].

We mention at this point that there exist studies expressing the four gluon vertex by means of three-point vertices which allows to dampen the scaling behaviour down to $O(n^3)$ [157–159]. This works with the introduction of additional tensor currents which require a significantly higher memory usage and the recursion looses elegance and simplicity. We did not investigate this approach any further.

Scaling behaviour of tree amplitudes for unitarity cuts: An m-point tree-level subamplitude that appears within a unitarity cut of topology $C_R = \{i_1, \ldots, i_R\}$ involves always two adjacent legs with (in general complex) on-shell loop momenta and $(m - 2)$ external legs. In Eq. (2.192), the most general dependence of the subamplitudes on flavour, additional Fourier parameters and on whether they contribute to the rational or to the cut-constructible part is given. To analyse the scaling behaviour of our implementation, we neglect for a moment all additional specifications and simply think of the subamplitude as a purely gluonic subamplitude. In terms of off-shell currents any subamplitude then reads

$$A_m^{\text{cut-tree}}(C_R, i_r) = A_m^{\text{tree-cut}}(-\ell_r, p_{i_r}, \ldots, p_{i_{(r+1)}-1}, \ell_{r+1})$$
$$= J(-\ell_r, i_r, \ldots, i_{(r+1)} - 1) \cdot \mathcal{P}^{-1} \cdot J(\ell_{r+1})\Big|_{P_{1,m}=0}$$

with $\mathcal{P}^{-1}$ the inverse propagator of the $(m - 1)$-point current and $P_{1,m} = -\ell_r + p_{i_r} + \ldots + p_{i_{(r+1)}-1} + \ell_{r+1}$. The multiplicity m of a subamplitude is related to the topology as $m = 2 + [(n + i_{(r+1)} - i_r) \mod n]$ with n being the multiplicity of the one-loop primitive amplitude. We observe that in terms of the Berends-Giele recursion, the off-shell currents that involve only external momenta are the same for all possible subamplitudes from different unitarity cuts, assuming fixed helicities and fixed order of external legs in the one-loop n-point primitive amplitude. To take this fact into account we compute before any integrand evaluation via products of subamplitudes first all possible off-shell currents that involve exclusively external legs in order to reuse them again for every unitarity cut. This means that we have to fill also the blank entries in the triangular current array. We refer to it as the extended current matrix. Strictly speaking the highest multiplicity of external currents that appears in massless QCD unitarity cuts are $(n - 2)$-point currents due to the lack of tadpoles and on-shell bubbles. We ignore these technical details and think for simplicity of the extended current matrix as an $n \times n$ matrix. The computation of the current $J(-\ell_r, i_r, \ldots, i_{(r+1)} - 1)$ is then nothing else than formally inserting an additional column between column $(i_r - 1)$ and i_r of length $(m - 1)$. For $i_{r+1} < i_r$ all currents are taken modulo n. We have already derived in Eq. (B.1) that adding a column of currents is a process of order m^3 which means that once the extended current matrix is filled, the subamplitudes that may appear within unitarity cuts of an n-point primitive are computed with an asymptotic scaling of $O(n^3)$. It is worth mentioning that the $O(n^3)$ behaviour for large n is on average over all contributing subamplitudes the same for pure gluon and mixed quark–gluon primitive amplitudes. However, if the external currents are attached to a pure quark loop, the asymptotic behaviour reduces to $O(n^2)$ due to the lack of four-point interactions.

Filling the blank entries of the array in Fig. B.1 remains a process of $O(n^4)$ as can be seen by direct computation:

$$\sum_{k=2}^{n} \sum_{m=k}^{n} \frac{1}{2} m(m-1) = \frac{1}{8}n^4 - \frac{1}{12}n^3 - \frac{1}{8}n^2 + \frac{1}{12}n. \tag{B.3}$$

The number of subamplitudes needed for the computation of one primitive amplitude depends on the number of possible cuts which scales due to the pentagon contributions asymptotically as $O(n^5)$, c.f. section 3.2, leading to a total asymptotic behaviour of $O(n^8)$. Hence the initialisation phase to fill up the extended current matrix as an $O(n^4)$ process will asymptotically be negligible. We checked in explicit runtime measurements that this is true also for low multiplicities, as expected. We emphasise that the outlined improvement leaves the form of the recursion equations unchanged and is special to the situation encountered for trees within unitarity cuts.

As a final remark we remember that helicities of the particles inside the loop are summed over, i.e. for any subamplitude with fixed external helicities, we have in total four possible combinations of the loop helicities: $(+,+)$, $(+,-)$, $(-,+)$ and $(-,-)$. Since the $(m-1)$-point off-shell current is independent of the helicity of the mth particle, we can use the same current at any time for two helicity combinations: $(+,\pm)$ and $(-,\pm)$. The on-shell contraction with the final current is a constant time operation. Hence exploiting this trivial "cache" roughly doubles the runtime of the integrand evaluation.

B.2. Extension of the Berends-Giele recursion to arbitrary quark lines

Quark-gluon tree-level primitive amplitudes can be computed with only tiny modifications using the recursion equations described in section 1.1. We introduce now a simple index system which is very specific for QCD computations with only quarks and gluons and no other flavours. It is employed within NJET both for tree-level and one-loop primitive amplitudes. All quark lines are assumed to have different flavour, the equal flavour case of the full amplitude can be obtained upon anti-symmetrisation. The arrangement of quark flavours within a flavour list must match an allowed topology of primitive amplitudes such that no fermion lines within a primitive amplitude ever cross. For example, $[q_1, g, q_2, \bar{q}_2, \bar{q}_1]$ would be a valid choice, while $[q_1, g, q_2, \bar{q}_1, \bar{q}_2]$ cannot be matched to any primitive amplitude.)

The generalised form of the recursion equations reads

$$\mathcal{J}_x(1,...,n) = \mathcal{P}_x(P_{1,n}) \left[\sum_{i=1}^{n-1} V_{3;x_1x_2;x} \, \mathcal{J}_{x_1}(1,...,i) \, \mathcal{J}_{x_2}(i+1,...,n) \right. \tag{B.4}$$

$$\left. + \sum_{i=1}^{n-2} \sum_{j=i+1}^{n-1} V_{4;x_1x_2x_3;x} \, \mathcal{J}_{x_1}(1,...,i) \, \mathcal{J}_{x_2}(i+1,...,j) \, \mathcal{J}_{x_3}(j+1,...,n) \right].$$

$\mathcal{J}_x$ is now a generalised colour ordered off-shell current with an off-shell leg of flavour x. This can be either a gluon or an (anti-)quark. Any Lorentz or Dirac indices are implicitly contained in $\mathcal{J}_x$. The same holds for the propagator $\mathcal{P}_x$ and the colour ordered three-point and four-point vertices V_3 and V_4 given in Eqs. (1.18), (1.19) and (1.28). The notation of the vertices is adapted to the numerical bottom-up approach: The first two (three) indices x_1, x_2, (x_3) connect in clockwise order matching flavours of lower point currents to result in the n-point current with common flavour x. Although there are only two different colour ordered quark–gluon vertices, in the numerical bottom-up approach we have to distinguish which flavours are sitting on the first two positions to determine the flavour of the third outgoing particle, i.e. the flavour of the off-shell current. Hence, there are effectively six quark–gluon vertices which correspond to six different production mechanisms. These possible processes are listed in the first column of table B.1. For example, the notation $q + g \leftarrow q$ means physically the production of a quark and a gluon from the third (incoming) quark. Since we choose all vertices outgoing, this corresponds to the process $q + g + \bar{q} \leftarrow 0$ and therefore to the vertex $V_{3;qg;\bar{q}}$. The vertices of all processes can be read off from the third column in table B.1.

The index system works now as follows: Every external particle's flavour within a primitive amplitude is specified by an integer number:

- Each gluon gets a flavour index 0.

- Each (outgoing) quark is labelled with a positive integer $i \in \{1, 2, \ldots, N_{q\bar{q}}\}$ with $N_{q\bar{q}}$ being the total number of quark lines. At every fermion line, the corresponding (outgoing) anti-quark is labelled with $-i$.

With a valid flavour and helicity list, the one-point currents are determined as polarisation vectors and spinors. With the conditions in the second column in Tab. B.1, currents of flavour x_1, x_2 (and x_3) are connected unambiguously with the correct vertex. $\wedge$ means the logical conjunction between two statements and $\vee$ the logical inclusive disjunction. The fourth column gives the flavour of the resulting off-shell current, though in the notation of an incoming particle. The reason for the opposite nomenclature is that the flavour of this off-shell current is automatically outgoing when it is used later on in the recursion again as lower point current. The last row of Tab. B.1 becomes relevant if none of the process condition matches. In this case the vertex function returns and proceeds with the next summand contributing to this distinct off-shell current. If no match occurs for all possible combinations, the current is assigned a nogo number X which is different from all quark flavour assignments ($X = 99999$ in NJET).

process	condition	vertex	current
$g + g \leftarrow g$	$x_1 + x_2 = 0 \ \wedge \ x_1 x_2 = 0$	$V_{3;gg;g}$	$x = 0$
$\bar{q} + q \leftarrow g$	$x_1 + x_2 = 0 \ \wedge \ x_1 x_2 < 0 \ \wedge \ x_1 < 0$	$V_{3;\bar{q}q;g}$	$x = 0$
$q + \bar{q} \leftarrow g$	$x_1 + x_2 = 0 \ \wedge \ x_1 x_2 < 0 \ \wedge \ x_1 > 0$	$V_{3;q\bar{q};g}$	$x = 0$
$\bar{q} + g \leftarrow \bar{q}$	$x_1 + x_2 = x_1 \ \wedge \ x_1 < 0$	$V_{3;\bar{q}g;q}$	$x = x_1$
$q + g \leftarrow q$	$x_1 + x_2 = x_1 \ \wedge \ x_1 > 0$	$V_{3;qg;\bar{q}}$	$x = x_1$
$g + \bar{q} \leftarrow \bar{q}$	$x_1 + x_2 = x_2 \ \wedge \ x_2 < 0$	$V_{3;g\bar{q};q}$	$x = x_2$
$g + q \leftarrow q$	$x_1 + x_2 = x_2 \ \wedge \ x_2 > 0$	$V_{3;gq;\bar{q}}$	$x = x_2$
$g + g + g \leftarrow g$	$|x_1| + |x_2| + |x_3| = 0$	$V_{4;ggg;g}$	$x = 0$
$\times$	no match $\vee \ x_1 = X \ \vee \ x_2 = X$	return	$(x = X)$

Table B.1.: The quark index system: x_1, x_2 (and x_3) are the flavour indices of the generalised currents on the right hand side of Eq. (B.4). The matching condition selects either unambiguously the correct vertex, or the vertex does not exist. Non existing currents are assigned a "nogo" index X different from all external flavour indices.

Algorithmically X is treated as an additional single flavour. A missing match occurs either if $x_1 = X$ or $x_2 = X$ or if quarks of unequal flavour should be connected.

In NJET , the currents are objects which include the momentum, the mass, the particle index and the actual numerical value of the current (a four-dimensional contravariant Lorentz vector or a four-component Dirac spinor). The recursion is then implemented with two general vertex functions **vertex3**$(\mathcal{J}_{x_1}, \mathcal{J}_{x_2}, \mathcal{J}_x)$ and **vertex4**$(\mathcal{J}_{x_1}, \mathcal{J}_{x_2}, \mathcal{J}_{x_3}, \mathcal{J}_x)$ which take as arguments currents and decide internally with help of the current indices, which vertex will be used. The interested reader may directly look into the implementation of the C++ class `Current` which computes all tree-level amplitudes within NJET . Further details to use the `Current` class as a matrix element generator can be found in the NJET documentation in Ref. [80].

The index system in conjunction with the recursive bottom-up approach presented in appendix B.1 works because in pure QCD computations, the flavour assignment of the new off-shell current from lower point currents is unique. However, if we had to include for example internal scalars (like in $\mathcal{N} = 4$ SYM) or a Z boson which do not appear as external particles, the assignment of currents would be ambiguous. An example of double assignments are the two processes $q + \bar{q} \leftarrow g$ and $q + \bar{q} \leftarrow s$ where s is a scalar. Recursive computations are still possible, but the system needs to be modified.

The problem of double assignment does not happen including the scalars that we need for the subamplitudes of the rational part. The reason for this is simply because the scalars are used *instead* of gluons and occur exclusively within the loop. In terms of subamplitudes that contribute to the integrand, these are external legs. The computation of subamplitudes for unitarity cuts is technically realised with modified vertex functions **vertex3**$_w(\mathcal{J}^*_{x_1}, \mathcal{J}_{x_2}, \mathcal{J}^*_x)$ and **vertex4**$_w(\mathcal{J}^*_{x_1}, \mathcal{J}_{x_2}, \mathcal{J}_{x_3}, \mathcal{J}^*_x)$ which carry an additional argument $w \in \{c, r\}$ that decides whether using gluons (c, present in the cut-constructible part) or scalars (r, present in the rational part). The star on the current arguments further indicates that the corresponding momenta are complex. The currents with real external momenta are all cached in the initialisation phase and need not be recomputed, c.f. the discussion at the end of appendix B.1. The vertices needed for the cut-constructible subamplitudes, are all contained in table B.1. The quark-index system for the rational subamplitudes stays the same, now the scalar carries index 0 instead of the gluon. The vertices, however, change according to the following replacement rules:

$$V_{3;gg;g} \to V_{3;sg;s} \qquad V_{4;ggg;g} \to V_{4;sgg;s}$$
$$V_{3;\bar{q}q;g} \to V_{3;\bar{q}q;s} \qquad V_{3;q\bar{q};g} \to V_{3;q\bar{q};s}$$
$$V_{3;g\bar{q};q} \to V_{3;s\bar{q};q} \qquad V_{3;gq;\bar{q}} \to V_{3;sq;\bar{q}}$$

The vertices $V_{3;sg;s}$, $V_{4;sgg;s}$ are given in Eqs. (2.93) – (2.94) and $V_{3;\bar{q}q;s}$, $V_{3;q\bar{q};s}$, $V_{3;s\bar{q};q}$, $V_{3;sq;\bar{q}}$ are given in Eqs. (2.101) – (2.102). The vertices $V_{3;\bar{q}q;g}$ and $V_{3;\bar{q}g;q}$ are the same for rational and cut-constructible subamplitudes.

C. Some Gram determinant relations

We derive in this appendix Eqs. (2.29) – (2.31). For this purpose, we write the projection operator onto the transverse space (2.28) in explicit determinant form.

$$
w_\mu{}^\nu = \frac{\delta^{k_1 k_2 \ldots k_{R-1} \nu}_{k_1 k_2 \ldots k_{R-1} \mu}}{\Delta(k_1, \ldots, k_{R-1})}
$$

$$
= \frac{1}{\Delta}
\begin{vmatrix}
k_1 k_1 & k_1 k_2 & \cdots & k_1 k_{R-1} & k_1^\nu \\
k_2 k_1 & k_2 k_2 & \cdots & k_2 k_{R-1} & k_2^\nu \\
\vdots & \vdots & & \vdots & \vdots \\
k_{R-1} k_1 & k_{R-1} k_2 & \cdots & k_{R-1} k_{R-1} & k_{R-1}^\nu \\
k_{1\mu} & k_{2\mu} & \cdots & k_{R-1\mu} & \delta_\mu^\nu
\end{vmatrix}
\tag{C.1}
$$

To see why Eq (2.29) is true, one can expand the determinant in the numerator of Eq. (C.1) via Laplace's formula:

$$
w_\mu{}^\mu = \frac{1}{\Delta}\left(\delta_\mu^\mu \Delta - k_{R-1\mu}
\begin{vmatrix}
k_1 k_1 & \cdots & k_1 k_{R-2} & k_{1\mu} \\
\vdots & & \vdots & \vdots \\
k_{R-1} k_1 & \cdots & k_{R-1} k_{R-2} & k_{R-1\mu}
\end{vmatrix}
\right.
$$

$$
\left. + k_{R-2\mu}
\begin{vmatrix}
k_1 k_1 & \cdots & k_1 k_{R-3} & k_1 k_{R-1} & k_{1\mu} \\
\vdots & & \vdots & \vdots & \vdots \\
k_1 R - 1 k_1 & \cdots & k_{R-1} k_{R-3} & k_{R-1} k_{R-1} & k_{R-1\mu}
\end{vmatrix}
- \cdots \right)
$$

Now, one simply contracts the vectors in front of the determinants with the last row, and finally interchanges the columns within the determinant such that one arrives at the known form of the $R - 1$ particle Gram determinant Δ. Interchanging the columns makes that all signs in front of Δ are equal. Thus, one gets

$$
w_\mu{}^\mu = \frac{1}{\Delta}(D\,\Delta - (R-1)\Delta) = D - R + 1 = D_T.
$$

Eqs. (2.30) are immediately fulfilled because a contraction with any vector k_i results in a second identical column and thus the determinant vanishes. Since the v_i are linear combinations of the k_i, the contraction with the v_i is zero as well. For the last identity,

Eq. (2.31), one writes

$$
w_\mu{}^\alpha w_\alpha{}^\nu = \frac{1}{\Delta^2} \delta^{k_1 k_2 \ldots k_{R-1} \alpha}_{k_1 k_2 \ldots k_{R-1} \mu} \delta^{k_1 k_2 \ldots k_{R-1} \nu}_{k_1 k_2 \ldots k_{R-1} \alpha}
$$

$$
= \frac{1}{\Delta^2} \left(\delta^\alpha_\mu \Delta - k^\alpha_{R-1} \det_{R-1}(k_\mu) + k^\alpha_{R-2} \det_{R-2}(k_\mu) - \ldots \right) \times
$$

$$
\times \begin{vmatrix}
k_1 k_1 & k_1 k_2 & \cdots & k_1 k_{R-1} & k_1^\nu \\
k_2 k_1 & k_2 k_2 & \cdots & k_2 k_{R-1} & k_2^\nu \\
\vdots & \vdots & & \vdots & \vdots \\
k_{R-1} k_1 & k_{R-1} k_2 & \cdots & k_{R-1} k_{R-1} & k_{R-1}^\nu \\
k_{1\alpha} & k_{2\alpha} & \cdots & k_{R-1\alpha} & \delta^\nu_\alpha
\end{vmatrix}
$$

where we have expanded the first determinant in the last row as in the previous example and $\det_{R-i}(k_\mu)$ is a shorthand notation for the corresponding subdeterminants. One sees that any contraction of k_i^α with the unexpanded determinant gives zero since two of the R rows are identical. The only non-vanishing contribution arises from the contraction with δ^α_μ. Hence, we get

$$
w_\mu{}^\alpha w_\alpha{}^\nu = \frac{1}{\Delta^2} \Delta \, \delta^\alpha_\mu \, \delta^{k_1 k_2 \ldots k_{R-1} \nu}_{k_1 k_2 \ldots k_{R-1} \alpha} = w_\mu{}^\nu
$$

and raising the μ index, we have

$$
w^{\mu\alpha} w_\alpha{}^\nu = w^{\mu\nu}.
$$

D. Alternative methods to disentangle the integral coefficients

D.1. The cut-constructible integral coefficient $b^{(0)}$

Since we are interested in $b^{(0)}$ exclusively and not in any other spurious terms, there exist clever choices of the triples $\vec{\alpha}_i = (\alpha_1, \alpha_2, \alpha_3)$ in Eq. (2.157) which allow us to evaluate the integrand less often than there are independent coefficients $b^{(i)}$. Such a clever choice are the vertices of platonic solids. These vertices are all equally separated on a three dimensional sphere, i.e. there is no point superior with respect to all others. Per constructionem, they automatically fulfil the unitarity constraints in Eq. (2.157). Of special interest are the tetrahedron, hexahedron (cube), and the octahedron because their number of vertices is less than the number of independent integrand terms $b^{(i)}$. The *tetrahedron* has four vertices which we can choose as our triplets $\vec{\alpha}_i$:

$$\vec{\alpha}_1 = \kappa\left(0, 0, 1\right) \qquad \vec{\alpha}_2 = \kappa\left(\frac{2\sqrt{2}}{3}, 0, -\frac{1}{3}\right) \qquad \vec{\alpha}_{3,4} = \kappa\left(-\frac{\sqrt{2}}{3}, \pm\sqrt{\frac{2}{3}}, -\frac{1}{3}\right). \qquad \text{(D.1)}$$

By explicitly plugging these numbers in Eq. (2.162), one sees that

$$b^{(0)} = \frac{1}{4}\sum_{j=1}^{4} \bar{b}_{i_1 i_2}^{[4]}(\ell(\vec{\alpha}_j)) \qquad \text{(D.2)}$$

In this approach there are only four integrand evaluations needed. The *octahedron* has six vertices which we choose as

$$\vec{\alpha}_{1,2} = (\pm\kappa, 0, 0) \qquad \vec{\alpha}_{3,4} = (0, \pm\kappa, 0) \qquad \vec{\alpha}_{5,6} = (0, 0, \pm\kappa) \qquad \text{(D.3)}$$

The integral coefficient can be computed with six function calls and reads

$$b^{(0)} = \frac{1}{6}\sum_{j=1}^{6} \bar{b}_{i_1 i_2}(\ell(\vec{\alpha}_j)). \qquad \text{(D.4)}$$

as can directly be verified by plugging these six triplets into Eq. (2.162). The *hexahedron* has eight vertices which we choose as our triplets $\vec{\alpha}_i$:

$$\vec{\alpha}_{1,2} = \frac{\kappa}{\sqrt{3}}(\pm 1, \pm 1, \pm 1) \qquad \vec{\alpha}_{3,4} = \frac{\kappa}{\sqrt{3}}(\mp 1, \pm 1, \pm 1)$$

$$\vec{\alpha}_{5,6} = \frac{\kappa}{\sqrt{3}}(\pm 1, \mp 1, \pm 1) \qquad \vec{\alpha}_{7,8} = \frac{\kappa}{\sqrt{3}}(\pm 1, \pm 1, \mp 1) \tag{D.5}$$

By explicitly plugging the numbers into Eq. (2.162), one sees that

$$b^{(0)} = \frac{1}{8} \sum_{j=1}^{8} \bar{b}_{i_1 i_2}(\ell(\vec{\alpha}_j)). \tag{D.6}$$

In this approach there are eight integrand evaluations needed. In terms of naive run-time performance, any of these three alternatives would be better than the method with discrete Fourier Transformation as described in section 2.5.3. Since we have not tested this sort of disentanglement on numerical accuracy, we cannot say if the speed up is worth the effort. In case it is worse and we have to switch to quadruple precision for a higher amount of phase space points, the overall performance can even be slower. We postpone this interesting investigation for a future release.

D.2. The rational integral coefficient $b^{(9)}$

With the findings of the previous section on the computation of $b^{(0)}$ in the cut-construct-ible case, it is straightforward to extend it to the rational integral coefficient $b^{(9)}$. The minor modification is that κ depends now on the free parameter μ^2. Hence any of the previous solutions in terms of platonic solids may be followed by a Fourier Projection on μ^2 according to

$$b^{(9)} = \frac{1}{q+1} \sum_{m=0}^{q} \frac{1}{r_0^2} \exp\left(-\frac{2\pi i}{q+1} m\right) \mathcal{P}_{\text{platonic}}[\bar{b}(\ell, \mu^2)] \tag{D.7}$$

where r_0 is the Fourier radius as described in section 2.6.2. The minimal number of integrand evaluations corresponds to the choice $q = 1$ leading to two modes in μ^2 and the tetrahedron as the platonic solid. All in all, these are eight integrand evaluations. The explicit formula reads

$$b^{(9)} = \frac{1}{8} \frac{1}{r_0^2} \sum_{\sigma=\pm} \sigma \sum_{j=1}^{4} \bar{b}(\ell_{\sigma,j}) \tag{D.8}$$

where $\ell_{\sigma,j}$ is the loop momentum with the explicit choices for $\vec{\alpha}_i$ in Eq. (D.1) and a σ-dependent κ

$$\kappa_\sigma = \sqrt{-(V_2^2 - \sigma r_0^2)}. \tag{D.9}$$

Bibliography

[1] T. Stelzer and W. Long, "Automatic generation of tree level helicity amplitudes," *Comput.Phys.Commun.* **81** (1994) 357–371, arXiv:hep-ph/9401258 [hep-ph].

[2] F. Krauss, R. Kuhn, and G. Soff, "AMEGIC++ 1.0: A Matrix element generator in C++," *JHEP* **0202** (2002) 044, arXiv:hep-ph/0109036 [hep-ph].

[3] T. Gleisberg and S. Hoeche, "Comix, a new matrix element generator," *JHEP* **0812** (2008) 039, arXiv:0808.3674 [hep-ph].

[4] A. Cafarella, C. G. Papadopoulos, and M. Worek, "Helac-Phegas: A Generator for all parton level processes," *Comput.Phys.Commun.* **180** (2009) 1941–1955, arXiv:arXiv:0710.2427 [hep-ph].

[5] M. L. Mangano, M. Moretti, F. Piccinini, R. Pittau, and A. D. Polosa, "ALPGEN, a generator for hard multiparton processes in hadronic collisions," *JHEP* **0307** (2003) 001, arXiv:hep-ph/0206293 [hep-ph].

[6] E. Boos *et al.*, "CompHEP 4.4: Automatic computations from Lagrangians to events," *Nucl.Instrum.Meth.* **A534** (2004) 250–259, arXiv:hep-ph/0403113 [hep-ph].

[7] M. Moretti, T. Ohl, and J. Reuter, "O'Mega: An Optimizing matrix element generator," arXiv:hep-ph/0102195 [hep-ph].

[8] S. Catani and M. Seymour, "The Dipole formalism for the calculation of QCD jet cross-sections at next-to-leading order," *Phys.Lett.* **B378** (1996) 287–301, arXiv:hep-ph/9602277 [hep-ph].

[9] S. Catani, S. Dittmaier, M. H. Seymour, and Z. Trocsanyi, "The Dipole formalism for next-to-leading order QCD calculations with massive partons," *Nucl.Phys.* **B627** (2002) 189–265, arXiv:hep-ph/0201036 [hep-ph].

[10] T. Gleisberg, S. Hoeche, F. Krauss, M. Schonherr, S. Schumann, *et al.*, "Event generation with SHERPA 1.1," *JHEP* **0902** (2009) 007, arXiv:0811.4622 [hep-ph].

[11] K. Hasegawa, S. Moch, and P. Uwer, "AutoDipole: Automated generation of dipole subtraction terms," *Comput.Phys.Commun.* **181** (2010) 1802–1817, arXiv:0911.4371 [hep-ph].

[12] M. H. Seymour and C. Tevlin, "TeVJet: A General framework for the calculation of jet observables in NLO QCD," arXiv:0803.2231 [hep-ph].

[13] R. Frederix, T. Gehrmann, and N. Greiner, "Automation of the Dipole Subtraction Method in MadGraph/MadEvent," *JHEP* **0809** (2008) 122, arXiv:0808.2128 [hep-ph].

[14] M. Czakon, C. Papadopoulos, and M. Worek, "Polarizing the Dipoles," *JHEP* **0908** (2009) 085, arXiv:0905.0883 [hep-ph].

[15] R. Frederix, S. Frixione, F. Maltoni, and T. Stelzer, "Automation of next-to-leading order computations in QCD: The FKS subtraction," *JHEP* **0910** (2009) 003, arXiv:0908.4272 [hep-ph].

[16] G. Passarino and M. Veltman, "One Loop Corrections for e+ e- Annihilation Into mu+ mu- in the Weinberg Model," *Nucl.Phys.* **B160** (1979) 151.

[17] W. van Neerven and J. Vermaseren, "LARGE LOOP INTEGRALS," *Phys.Lett.* **B137** (1984) 241.

[18] Z. Bern, L. J. Dixon, and D. A. Kosower, "Dimensionally regulated one loop integrals," *Phys.Lett.* **B302** (1993) 299–308, arXiv:hep-ph/9212308 [hep-ph].

[19] Z. Bern, L. J. Dixon, and D. A. Kosower, "Dimensionally regulated pentagon integrals," *Nucl.Phys.* **B412** (1994) 751–816, arXiv:hep-ph/9306240 [hep-ph].

[20] A. I. Davydychev, "A Simple formula for reducing Feynman diagrams to scalar integrals," *Phys.Lett.* **B263** (1991) 107–111.

[21] O. Tarasov, "Connection between Feynman integrals having different values of the space-time dimension," *Phys.Rev.* **D54** (1996) 6479–6490, arXiv:hep-th/9606018 [hep-th].

[22] J. Fleischer, F. Jegerlehner, and O. Tarasov, "Algebraic reduction of one loop Feynman graph amplitudes," *Nucl.Phys.* **B566** (2000) 423–440, arXiv:hep-ph/9907327 [hep-ph].

[23] T. Binoth, J. Guillet, and G. Heinrich, "Reduction formalism for dimensionally regulated one loop N point integrals," *Nucl.Phys.* **B572** (2000) 361–386, arXiv:hep-ph/9911342 [hep-ph].

[24] T. Binoth, J. P. Guillet, G. Heinrich, E. Pilon, and C. Schubert, "An Algebraic/numerical formalism for one-loop multi-leg amplitudes," *JHEP* **0510** (2005) 015, arXiv:hep-ph/0504267 [hep-ph].

[25] A. Denner and S. Dittmaier, "Reduction of one loop tensor five point integrals," *Nucl.Phys.* **B658** (2003) 175–202, arXiv:hep-ph/0212259 [hep-ph].

[26] A. Denner and S. Dittmaier, "Reduction schemes for one-loop tensor integrals," *Nucl.Phys.* **B734** (2006) 62–115, arXiv:hep-ph/0509141 [hep-ph].

[27] T. Diakonidis, J. Fleischer, J. Gluza, K. Kajda, T. Riemann, *et al.*, "A Complete reduction of one-loop tensor 5 and 6-point integrals," *Phys.Rev.* **D80** (2009) 036003, arXiv:0812.2134 [hep-ph].

[28] C. Berger, Z. Bern, L. J. Dixon, F. Febres Cordero, D. Forde, *et al.*, "Precise Predictions for W + 3 Jet Production at Hadron Colliders," *Phys.Rev.Lett.* **102** (2009) 222001, arXiv:0902.2760 [hep-ph].

[29] C. Berger, Z. Bern, L. J. Dixon, F. Febres Cordero, D. Forde, *et al.*, "Next-to-Leading Order QCD Predictions for W+3-Jet Distributions at Hadron Colliders," *Phys.Rev.* **D80** (2009) 074036, arXiv:0907.1984 [hep-ph].

[30] C. Berger, Z. Bern, L. J. Dixon, F. Febres Cordero, D. Forde, *et al.*, "Next-to-Leading Order QCD Predictions for Z, γ^*+3-Jet Distributions at the Tevatron," *Phys.Rev.* **D82** (2010) 074002, arXiv:1004.1659 [hep-ph].

[31] A. Bredenstein, A. Denner, S. Dittmaier, and S. Pozzorini, "NLO QCD corrections to pp → t anti-t b anti-b + X at the LHC," *Phys.Rev.Lett.* **103** (2009) 012002, arXiv:0905.0110 [hep-ph].

[32] A. Bredenstein, A. Denner, S. Dittmaier, and S. Pozzorini, "NLO QCD Corrections to Top Anti-Top Bottom Anti-Bottom Production at the LHC: 2. full hadronic results," *JHEP* **1003** (2010) 021, arXiv:1001.4006 [hep-ph].

[33] G. Bevilacqua, M. Czakon, C. Papadopoulos, R. Pittau, and M. Worek, "Assault on the NLO Wishlist: → t anti-t b anti-b," *JHEP* **0909** (2009) 109, arXiv:0907.4723 [hep-ph].

[34] F. Campanario, C. Englert, M. Rauch, and D. Zeppenfeld, "Precise predictions for W γ γ +jet production at hadron colliders," *Phys.Lett.* **B704** (2011) 515–519, arXiv:1106.4009 [hep-ph].

[35] G. Bevilacqua, M. Czakon, C. Papadopoulos, and M. Worek, "Dominant QCD Backgrounds in Higgs Boson Analyses at the LHC: A Study of pp → t anti-t + 2 jets at Next-To-Leading Order," *Phys.Rev.Lett.* **104** (2010) 162002, arXiv:1002.4009 [hep-ph].

[36] A. Denner, S. Dittmaier, S. Kallweit, and S. Pozzorini, "NLO QCD corrections to WWbb production at hadron colliders," *Phys.Rev.Lett.* **106** (2011) 052001, arXiv:1012.3975 [hep-ph].

[37] G. Bevilacqua, M. Czakon, A. van Hameren, C. G. Papadopoulos, and M. Worek, "Complete off-shell effects in top quark pair hadroproduction with leptonic decay at next-to-leading order," *JHEP* **1102** (2011) 083, arXiv:1012.4230 [hep-ph].

[38] G. Bevilacqua, M. Czakon, C. Papadopoulos, and M. Worek, "Hadronic top-quark pair production in association with two jets at Next-to-Leading Order QCD," *Phys.Rev.* **D84** (2011) 114017, arXiv:1108.2851 [hep-ph].

[39] A. Denner, S. Dittmaier, S. Kallweit, and S. Pozzorini, "NLO QCD corrections to off-shell top-antitop production with leptonic decays at hadron colliders," *JHEP* **1210** (2012) 110, arXiv:1207.5018 [hep-ph].

[40] G. Bevilacqua and M. Worek, "Constraining BSM Physics at the LHC: Four top final states with NLO accuracy in perturbative QCD," *JHEP* **1207** (2012) 111, arXiv:1206.3064 [hep-ph].

[41] T. Binoth, N. Greiner, A. Guffanti, J. Reuter, J.-P. Guillet, *et al.*, "Next-to-leading order QCD corrections to pp → b anti-b b anti-b + X at the LHC: the quark induced case," *Phys.Lett.* **B685** (2010) 293–296, arXiv:0910.4379 [hep-ph].

[42] N. Greiner, A. Guffanti, T. Reiter, and J. Reuter, "NLO QCD corrections to the production of two bottom-antibottom pairs at the LHC," *Phys.Rev.Lett.* **107** (2011) 102002, arXiv:1105.3624 [hep-ph].

[43] N. Greiner, G. Heinrich, P. Mastrolia, G. Ossola, T. Reiter, *et al.*, "NLO QCD corrections to the production of W+ W- plus two jets at the LHC," *Phys.Lett.* **B713** (2012) 277–283, arXiv:1202.6004 [hep-ph].

[44] K. Melnikov and G. Zanderighi, "W+3 jet production at the LHC as a signal or background," *Phys.Rev.* **D81** (2010) 074025, arXiv:0910.3671 [hep-ph].

[45] R. K. Ellis, K. Melnikov, and G. Zanderighi, "W+3 jet production at the Tevatron," *Phys.Rev.* **D80** (2009) 094002, arXiv:0906.1445 [hep-ph].

[46] T. Melia, K. Melnikov, R. Rontsch, and G. Zanderighi, "NLO QCD corrections for W^+W^- pair production in association with two jets at hadron colliders," *Phys.Rev.* **D83** (2011) 114043, arXiv:1104.2327 [hep-ph].

[47] T. Melia, K. Melnikov, R. Rontsch, and G. Zanderighi, "Next-to-leading order QCD predictions for W^+W^+jj production at the LHC," *JHEP* **1012** (2010) 053, arXiv:1007.5313 [hep-ph].

[48] A. Denner, L. Hosekova, and S. Kallweit, "NLO QCD corrections to W+ W+ jj production in vector-boson fusion at the LHC," *Phys.Rev.* **D86** (2012) 114014, arXiv:1209.2389 [hep-ph].

[49] C. Berger, Z. Bern, L. J. Dixon, F. Febres Cordero, D. Forde, *et al.*, "Precise Predictions for W + 4 Jet Production at the Large Hadron Collider," *Phys.Rev.Lett.* **106** (2011) 092001, arXiv:1009.2338 [hep-ph].

[50] H. Ita, Z. Bern, L. Dixon, F. Febres Cordero, D. Kosower, *et al.*, "Precise Predictions for Z + 4 Jets at Hadron Colliders," *Phys.Rev.* **D85** (2012) 031501, arXiv:1108.2229 [hep-ph].

[51] R. Frederix, S. Frixione, K. Melnikov, and G. Zanderighi, "NLO QCD corrections to five-jet production at LEP and the extraction of $\alpha_s(M_Z)$," *JHEP* **1011** (2010) 050, arXiv:1008.5313 [hep-ph].

[52] S. Becker, D. Goetz, C. Reuschle, C. Schwan, and S. Weinzierl, "NLO results for five, six and seven jets in electron-positron annihilation," *Phys.Rev.Lett.* **108** (2012) 032005, arXiv:1111.1733 [hep-ph].

[53] Z. Bern, L. Dixon, F. F. Cordero, S. Hoeche, H. Ita, *et al.*, "Next-to-Leading Order W + 5-Jet Production at the LHC," arXiv:1304.1253 [hep-ph].

[54] Z. Bern, L. J. Dixon, D. C. Dunbar, and D. A. Kosower, "One loop n point gauge theory amplitudes, unitarity and collinear limits," *Nucl.Phys.* **B425** (1994) 217–260, arXiv:hep-ph/9403226 [hep-ph].

[55] Z. Bern, L. J. Dixon, D. C. Dunbar, and D. A. Kosower, "Fusing gauge theory tree amplitudes into loop amplitudes," *Nucl.Phys.* **B435** (1995) 59–101, arXiv:hep-ph/9409265 [hep-ph].

[56] R. Cutkosky, "Singularities and discontinuities of Feynman amplitudes," *J.Math.Phys.* **1** (1960) 429–433.

[57] R. Britto, F. Cachazo, and B. Feng, "Generalized unitarity and one-loop amplitudes in N=4 super-Yang-Mills," *Nucl.Phys.* **B725** (2005) 275–305, arXiv:hep-th/0412103 [hep-th].

[58] R. Britto, E. Buchbinder, F. Cachazo, and B. Feng, "One-loop amplitudes of gluons in SQCD," *Phys.Rev.* **D72** (2005) 065012, arXiv:hep-ph/0503132 [hep-ph].

[59] R. Britto, B. Feng, and P. Mastrolia, "The Cut-constructible part of QCD amplitudes," *Phys.Rev.* **D73** (2006) 105004, arXiv:hep-ph/0602178 [hep-ph].

[60] D. Forde, "Direct extraction of one-loop integral coefficients," *Phys.Rev.* **D75** (2007) 125019, arXiv:0704.1835 [hep-ph].

[61] Z. Bern, L. J. Dixon, and D. A. Kosower, "One loop amplitudes for e+ e- to four partons," *Nucl.Phys.* **B513** (1998) 3–86, arXiv:hep-ph/9708239 [hep-ph].

[62] Z. Bern and A. Morgan, "Massive loop amplitudes from unitarity," *Nucl.Phys.* **B467** (1996) 479–509, arXiv:hep-ph/9511336 [hep-ph].

[63] Z. Bern, L. J. Dixon, and D. A. Kosower, "Bootstrapping multi-parton loop amplitudes in QCD," *Phys.Rev.* **D73** (2006) 065013, arXiv:hep-ph/0507005 [hep-ph].

[64] C. F. Berger, Z. Bern, L. J. Dixon, D. Forde, and D. A. Kosower, "Bootstrapping One-Loop QCD Amplitudes with General Helicities," *Phys.Rev.* **D74** (2006) 036009, arXiv:hep-ph/0604195 [hep-ph].

[65] C. Anastasiou, R. Britto, B. Feng, Z. Kunszt, and P. Mastrolia, "D-dimensional unitarity cut method," *Phys.Lett.* **B645** (2007) 213–216, arXiv:hep-ph/0609191 [hep-ph].

[66] C. Anastasiou, R. Britto, B. Feng, Z. Kunszt, and P. Mastrolia, "Unitarity cuts and Reduction to master integrals in d dimensions for one-loop amplitudes," *JHEP* **0703** (2007) 111, arXiv:hep-ph/0612277 [hep-ph].

[67] S. Badger, "Direct Extraction Of One Loop Rational Terms," *JHEP* **0901** (2009) 049, arXiv:0806.4600 [hep-ph].

[68] G. Ossola, C. G. Papadopoulos, and R. Pittau, "Reducing full one-loop amplitudes to scalar integrals at the integrand level," *Nucl.Phys.* **B763** (2007) 147–169, arXiv:hep-ph/0609007 [hep-ph].

[69] G. Ossola, C. G. Papadopoulos, and R. Pittau, "On the Rational Terms of the one-loop amplitudes," *JHEP* **0805** (2008) 004, arXiv:0802.1876 [hep-ph].

[70] W. T. Giele, Z. Kunszt, and K. Melnikov, "Full one-loop amplitudes from tree amplitudes," *JHEP* **0804** (2008) 049, arXiv:0801.2237 [hep-ph].

[71] R. K. Ellis, W. T. Giele, Z. Kunszt, and K. Melnikov, "Masses, fermions and generalized D-dimensional unitarity," *Nucl.Phys.* **B822** (2009) 270–282, arXiv:0806.3467 [hep-ph].

[72] R. K. Ellis, W. Giele, and Z. Kunszt, "A Numerical Unitarity Formalism for Evaluating One-Loop Amplitudes," *JHEP* **0803** (2008) 003, arXiv:0708.2398 [hep-ph].

[73] H. Ita, "Susy Theories and QCD: Numerical Approaches," *J.Phys.* **A44** (2011) 454005, arXiv:1109.6527 [hep-th].

[74] Z. Bern, L. J. Dixon, and D. A. Kosower, "One loop corrections to two quark three gluon amplitudes," *Nucl.Phys.* **B437** (1995) 259–304, arXiv:hep-ph/9409393 [hep-ph].

[75] R. K. Ellis, Z. Kunszt, K. Melnikov, and G. Zanderighi, "One-loop calculations in quantum field theory: from Feynman diagrams to unitarity cuts," *Phys.Rept.* **518** (2012) 141–250, arXiv:1105.4319 [hep-ph].

[76] H. Ita and K. Ozeren, "Colour Decompositions of Multi-quark One-loop QCD Amplitudes," *JHEP* **1202** (2012) 118, arXiv:1111.4193 [hep-ph].

[77] S. Badger, B. Biedermann, and P. Uwer, "NGluon: A Package to Calculate One-loop Multi-gluon Amplitudes," *Comput.Phys.Commun.* **182** (2011) 1674–1692, arXiv:1011.2900 [hep-ph].

[78] S. Badger, B. Biedermann, and P. Uwer, "Numerical Evaluation of One-Loop QCD Amplitudes," *J.Phys.Conf.Ser.* **368** (2012) 012055, arXiv:1112.0412 [hep-ph].

[79] S. Badger, B. Biedermann, and P. Uwer, "One-Loop Amplitudes for Multi-Jet Production at Hadron Colliders," *PoS* **RADCOR2011** (2011) 020, arXiv:1201.1187 [hep-ph].

[80] S. Badger, B. Biedermann, P. Uwer, and V. Yundin, "Numerical evaluation of virtual corrections to multi-jet production in massless QCD," *Comput.Phys.Commun.* **184** (2013) 1981–1998, arXiv:1209.0100 [hep-ph].

[81] S. Actis, A. Denner, L. Hofer, A. Scharf, and S. Uccirati, "Recursive generation of one-loop amplitudes in the Standard Model," *JHEP* **1304** (2013) 037, arXiv:1211.6316 [hep-ph].

[82] F. Cascioli, P. Maierhofer, and S. Pozzorini, "Scattering Amplitudes with Open Loops," *Phys.Rev.Lett.* **108** (2012) 111601, arXiv:1111.5206 [hep-ph].

[83] G. Bevilacqua, M. Czakon, M. Garzelli, A. van Hameren, A. Kardos, *et al.*, "HELAC-NLO," *Comput.Phys.Commun.* **184** (2013) 986–997, arXiv:1110.1499 [hep-ph].

[84] C. Berger, Z. Bern, L. Dixon, F. Febres Cordero, D. Forde, *et al.*, "An Automated Implementation of On-Shell Methods for One-Loop Amplitudes," *Phys.Rev.* **D78** (2008) 036003, arXiv:0803.4180 [hep-ph].

[85] G. Ossola, C. G. Papadopoulos, and R. Pittau, "CutTools: A Program implementing the OPP reduction method to compute one-loop amplitudes," *JHEP* **0803** (2008) 042, arXiv:0711.3596 [hep-ph].

[86] P. Mastrolia, G. Ossola, T. Reiter, and F. Tramontano, "Scattering AMplitudes from Unitarity-based Reduction Algorithm at the Integrand-level," *JHEP* **1008** (2010) 080, arXiv:1006.0710 [hep-ph].

[87] W. Giele and G. Zanderighi, "On the Numerical Evaluation of One-Loop Amplitudes: The Gluonic Case," *JHEP* **0806** (2008) 038, arXiv:0805.2152 [hep-ph].

[88] V. Hirschi, R. Frederix, S. Frixione, M. V. Garzelli, F. Maltoni, *et al.*, "Automation of one-loop QCD corrections," *JHEP* **1105** (2011) 044, arXiv:1103.0621 [hep-ph].

[89] G. Cullen, N. Greiner, G. Heinrich, G. Luisoni, P. Mastrolia, *et al.*, "Automated One-Loop Calculations with GoSam," *Eur.Phys.J.* **C72** (2012) 1889, arXiv:1111.2034 [hep-ph].

[90] S. Becker, C. Reuschle, and S. Weinzierl, "Efficiency Improvements for the Numerical Computation of NLO Corrections," *JHEP* **1207** (2012) 090, arXiv:1205.2096 [hep-ph].

[91] S. Badger, B. Biedermann, P. Uwer, and V. Yundin, "NLO QCD corrections to multi-jet production at the LHC with a centre-of-mass energy of $\sqrt{s} = 8$ TeV," *Phys.Lett.* **B718** (2013) 965–978, arXiv:1209.0098 [hep-ph].

[92] S. D. Ellis, Z. Kunszt, and D. E. Soper, "Two jet production in hadron collisions at order α_s^3 in QCD," *Phys.Rev.Lett.* **69** (1992) 1496–1499.

[93] W. Giele, E. N. Glover, and D. A. Kosower, "Higher order corrections to jet cross-sections in hadron colliders," *Nucl.Phys.* **B403** (1993) 633–670, arXiv:hep-ph/9302225 [hep-ph].

[94] Z. Nagy, "Three jet cross-sections in hadron hadron collisions at next-to-leading order," *Phys.Rev.Lett.* **88** (2002) 122003, arXiv:hep-ph/0110315 [hep-ph].

[95] Z. Nagy, "Next-to-leading order calculation of three jet observables in hadron hadron collision," *Phys.Rev.* **D68** (2003) 094002, arXiv:hep-ph/0307268 [hep-ph].

[96] W. B. Kilgore and W. Giele, "Next-to-leading order gluonic three jet production at hadron colliders," *Phys.Rev.* **D55** (1997) 7183–7190, arXiv:hep-ph/9610433 [hep-ph].

[97] Z. Trocsanyi, "Three jet cross-section in hadron collisions at next-to-leading order: Pure gluon processes," *Phys.Rev.Lett.* **77** (1996) 2182–2185, arXiv:hep-ph/9610499 [hep-ph].

[98] Z. Bern, G. Diana, L. Dixon, F. Febres Cordero, S. Hoeche, *et al.*, "Four-Jet Production at the Large Hadron Collider at Next-to-Leading Order in QCD," *Phys.Rev.Lett.* **109** (2012) 042001, arXiv:1112.3940 [hep-ph].

[99] F. A. Berends and W. Giele, "Recursive Calculations for Processes with n Gluons," *Nucl.Phys.* **B306** (1988) 759.

[100] M. L. Mangano and S. J. Parke, "Multiparton amplitudes in gauge theories," *Phys.Rept.* **200** (1991) 301–367, arXiv:hep-th/0509223 [hep-th].

[101] L. J. Dixon, "Calculating scattering amplitudes efficiently," arXiv:hep-ph/9601359 [hep-ph].

[102] R. Britto, F. Cachazo, B. Feng, and E. Witten, "Direct proof of tree-level recursion relation in Yang-Mills theory," *Phys.Rev.Lett.* **94** (2005) 181602, arXiv:hep-th/0501052 [hep-th].

[103] R. Roiban, M. Spradlin, and A. Volovich, "Scattering amplitudes in gauge theories: progress and outlook," *Journal of Physics A: Mathematical and Theoretical* **44** no. 45, (2011) 450301.

[104] B. Feng and M. Luo, "An Introduction to On-shell Recursion Relations," arXiv:1111.5759 [hep-th].

[105] R. Kleiss and H. Kuijf, "Multigluon cross-sections and five jet production at hadron colliders," *Nucl.Phys.* **B312** (1989) 616.

[106] V. Del Duca, L. J. Dixon, and F. Maltoni, "New color decompositions for gauge amplitudes at tree and loop level," *Nucl.Phys.* **B571** (2000) 51–70, arXiv:hep-ph/9910563 [hep-ph].

[107] Z. Bern, J. Carrasco, and H. Johansson, "New Relations for Gauge-Theory Amplitudes," *Phys.Rev.* **D78** (2008) 085011, arXiv:0805.3993 [hep-ph].

[108] F. Maltoni, K. Paul, T. Stelzer, and S. Willenbrock, "Color flow decomposition of QCD amplitudes," *Phys.Rev.* **D67** (2003) 014026, arXiv:hep-ph/0209271 [hep-ph].

[109] S. Weinzierl, "Automated computation of spin- and colour-correlated Born matrix elements," *Eur.Phys.J.* **C45** (2006) 745–757, arXiv:hep-ph/0510157 [hep-ph].

[110] W. Kilian, T. Ohl, J. Reuter, and C. Speckner, "QCD in the Color-Flow Representation," *JHEP* **1210** (2012) 022, arXiv:1206.3700 [hep-ph].

[111] L. Brink, J. H. Schwarz, and J. Scherk, "Supersymmetric Yang-Mills Theories," *Nucl.Phys.* **B121** (1977) 77.

[112] L. J. Dixon, J. M. Henn, J. Plefka, and T. Schuster, "All tree-level amplitudes in massless QCD," *JHEP* **1101** (2011) 035, arXiv:1010.3991 [hep-ph].

[113] S. Badger, B. Biedermann, L. Hackl, J. Plefka, T. Schuster, *et al.*, "Comparing efficient computation methods for massless QCD tree amplitudes: Closed Analytic Formulae versus Berends-Giele Recursion," *Phys. Rev.* **D87** (2013) 034011, arXiv:1206.2381 [hep-ph].

[114] P. Nogueira, "Automatic Feynman graph generation," *J.Comput.Phys.* **105** (1993) 279–289.

[115] Z. Bern and D. A. Kosower, "Color decomposition of one loop amplitudes in gauge theories," *Nucl.Phys.* **B362** (1991) 389–448.

[116] Z. Bern, L. J. Dixon, and D. A. Kosower, "One loop corrections to five gluon amplitudes," *Phys.Rev.Lett.* **70** (1993) 2677–2680, arXiv:hep-ph/9302280 [hep-ph].

[117] G. van Oldenborgh, "FF: A Package to evaluate one loop Feynman diagrams," *Comput.Phys.Commun.* **66** (1991) 1–15.

[118] R. K. Ellis and G. Zanderighi, "Scalar one-loop integrals for QCD," *JHEP* **0802** (2008) 002, arXiv:0712.1851 [hep-ph].

[119] T. Hahn and M. Perez-Victoria, "Automatized one loop calculations in four-dimensions and D-dimensions," *Comput.Phys.Commun.* **118** (1999) 153–165, arXiv:hep-ph/9807565 [hep-ph].

[120] A. van Hameren, "OneLOop: For the evaluation of one-loop scalar functions," *Comput.Phys.Commun.* **182** (2011) 2427–2438, arXiv:1007.4716 [hep-ph].

[121] T. Binoth, J.-P. Guillet, G. Heinrich, E. Pilon, and T. Reiter, "Golem95: A Numerical program to calculate one-loop tensor integrals with up to six external legs," *Comput.Phys.Commun.* **180** (2009) 2317–2330, arXiv:0810.0992 [hep-ph].

[122] G. Cullen, J. P. Guillet, G. Heinrich, T. Kleinschmidt, E. Pilon, *et al.*, "Golem95C: A library for one-loop integrals with complex masses," *Comput.Phys.Commun.* **182** (2011) 2276–2284, arXiv:1101.5595 [hep-ph].

[123] A. Bredenstein, A. Denner, S. Dittmaier, and S. Pozzorini, "NLO QCD corrections to t anti-t b anti-b production at the LHC: 1. Quark-antiquark annihilation," *JHEP* **0808** (2008) 108, arXiv:0807.1248 [hep-ph].

[124] S. Mandelstam, "Analytic properties of transition amplitudes in perturbation theory," *Phys.Rev.* **115** (1959) 1741–1751.

[125] L. Landau, "On analytic properties of vertex parts in quantum field theory," *Nucl.Phys.* **13** (1959) 181–192.

[126] R. Eden, P. Landshoff, D. Olive, and J. Polkinghorne, *The Analytic S-Matrix*. Camebridge University Press, 1966.

[127] Z. Bern and D. A. Kosower, "The Computation of loop amplitudes in gauge theories," *Nucl.Phys.* **B379** (1992) 451–561.

[128] Z. Kunszt, A. Signer, and Z. Trocsanyi, "One loop helicity amplitudes for all 2 →
2 processes in QCD and N=1 supersymmetric Yang-Mills theory," *Nucl.Phys.*
B411 (1994) 397–442, arXiv:hep-ph/9305239 [hep-ph].

[129] S. Catani, S. Dittmaier, and Z. Trocsanyi, "One loop singular behavior of QCD
and SUSY QCD amplitudes with massive partons," *Phys.Lett.* **B500** (2001)
149–160, arXiv:hep-ph/0011222 [hep-ph].

[130] V. Borodulin, R. Rogalev, and S. Slabospitsky, "CORE: COmpendium of
RElations: Version 2.1," arXiv:hep-ph/9507456 [hep-ph].

[131] K. Melnikov and M. Schulze, "NLO QCD corrections to top quark pair
production in association with one hard jet at hadron colliders," *Nucl.Phys.* **B840**
(2010) 129–159, arXiv:1004.3284 [hep-ph].

[132] T. Muta, "Foundations of quantum chromodynamics: An Introduction to
perturbative methods in gauge theories," *World Sci.Lect.Notes Phys.* **5** (1987)
1–409.

[133] Z. Bern, L. J. Dixon, and D. A. Kosower, "Progress in one loop QCD
computations," *Ann.Rev.Nucl.Part.Sci.* **46** (1996) 109–148,
arXiv:hep-ph/9602280 [hep-ph].

[134] L. Abbott, "The Background Field Method Beyond One Loop," *Nucl.Phys.* **B185**
(1981) 189.

[135] G. 't Hooft and M. Veltman, "Regularization and Renormalization of Gauge
Fields," *Nucl.Phys.* **B44** (1972) 189–213.

[136] W. Press, S. Teukolsky, W. Vetterling, and B. Flannery, *Numerical recipes in C —
The Art of Scientific Computing.* Camebridge University Press, 1992.

[137] P. Mastrolia, G. Ossola, C. Papadopoulos, and R. Pittau, "Optimizing the
Reduction of One-Loop Amplitudes," *JHEP* **0806** (2008) 030,
arXiv:0803.3964 [hep-ph].

[138] L. J. Guibas and R. Sedgewick, "A dichromatic framework for balanced trees,"
in *19th Annual Symposium on Foundations of Computer Science*, pp. 8–21. Oct.,
1978.

[139] Y. Hida, X. S. Li, and D. H. Bailey.
http://crd.lbl.gov/~dhbailey/mpdist/, 2010.

[140] D. Forde and D. A. Kosower, "All-multiplicity one-loop corrections to MHV
amplitudes in QCD," *Phys.Rev.* **D73** (2006) 061701, arXiv:hep-ph/0509358
[hep-ph].

[141] Z. Bern, L. J. Dixon, and D. A. Kosower, "The last of the finite loop amplitudes in QCD," *Phys.Rev.* **D72** (2005) 125003, arXiv:hep-ph/0505055 [hep-ph].

[142] E. Byckling and K. Kajantie, *Particle kinematics.* Wiley, 1973.

[143] R. Kleiss, W. J. Stirling, and S. Ellis, "A new Monte Carlo treatment of multiparticle phase space at high-energies," *Comput.Phys.Commun.* **40** (1986) 359.

[144] Z. Bern, L. J. Dixon, and D. A. Kosower, "On-shell recurrence relations for one-loop QCD amplitudes," *Phys.Rev.* **D71** (2005) 105013, arXiv:hep-th/0501240 [hep-th].

[145] **JADE Collaboration** Collaboration, W. Bartel *et al.*, "Experimental Studies on Multi-Jet Production in e+ e- Annihilation at PETRA Energies," *Z.Phys.* **C33** (1986) 23.

[146] **JADE Collaboration** Collaboration, S. Bethke *et al.*, "Experimental Investigation of the Energy Dependence of the Strong Coupling Strength," *Phys.Lett.* **B213** (1988) 235.

[147] A. van Hameren, "Multi-gluon one-loop amplitudes using tensor integrals," *JHEP* **0907** (2009) 088, arXiv:0905.1005 [hep-ph].

[148] R. K. Ellis and J. Sexton, "QCD Radiative Corrections to Parton Parton Scattering," *Nucl.Phys.* **B269** (1986) 445.

[149] T. Binoth, F. Boudjema, G. Dissertori, A. Lazopoulos, A. Denner, *et al.*, "A Proposal for a standard interface between Monte Carlo tools and one-loop programs," *Comput.Phys.Commun.* **181** (2010) 1612–1622, arXiv:1001.1307 [hep-ph].

[150] A. Martin, W. Stirling, R. Thorne, and G. Watt, "Parton distributions for the LHC," *Eur.Phys.J.* **C63** (2009) 189–285, arXiv:0901.0002 [hep-ph].

[151] **Particle Data Group** Collaboration, J. Beringer *et al.*, "Review of particle physics," *Phys. Rev. D* **86** (Jul, 2012) 010001.

[152] M. Cacciari, G. P. Salam, and G. Soyez, "The Anti-k(t) jet clustering algorithm," *JHEP* **0804** (2008) 063, arXiv:0802.1189 [hep-ph].

[153] M. Cacciari, G. P. Salam, and G. Soyez, "FastJet user manual," *Eur.Phys.J.* **C72** (2012) 1896, arXiv:1111.6097 [hep-ph].

[154] **ATLAS Collaboration** Collaboration, G. Aad *et al.*, "Measurement of multi-jet cross sections in proton-proton collisions at a 7 TeV center-of-mass energy," *Eur.Phys.J.* **C71** (2011) 1763, arXiv:1107.2092 [hep-ex].

[155] H. K. Dreiner, H. E. Haber, and S. P. Martin, "Two-component spinor techniques and Feynman rules for quantum field theory and supersymmetry," *Phys.Rept.* **494** (2010) 1–196, arXiv:0812.1594 [hep-ph].

[156] C. Schwinn and S. Weinzierl, "On-shell recursion relations for all Born QCD amplitudes," *JHEP* **0704** (2007) 072, arXiv:hep-ph/0703021 [HEP-PH].

[157] C. Duhr, S. Hoeche, and F. Maltoni, "Color-dressed recursive relations for multi-parton amplitudes," *JHEP* **0608** (2006) 062, arXiv:hep-ph/0607057 [hep-ph].

[158] W. Giele, G. Stavenga, and J.-C. Winter, "Thread-Scalable Evaluation of Multi-Jet Observables," *Eur.Phys.J.* **C71** (2011) 1703, arXiv:1002.3446 [hep-ph].

[159] P. Draggiotis, R. H. Kleiss, and C. G. Papadopoulos, "On the computation of multigluon amplitudes," *Phys.Lett.* **B439** (1998) 157–164, arXiv:hep-ph/9807207 [hep-ph].

List of Figures

List of Tables

Danksagung

Mein größter Dank gilt meinem Doktorvater Peter Uwer für die hervorragende Betreuung und dafür, mir diese spannende Zeit der Doktorarbeit ermöglicht zu haben. In den unzähligen Fachdiskussionen hatte er stets ein offenes Ohr, gute Ratschläge und verstand es wirklich sehr gut, mich auch während Durststrecken immer wieder zu motivieren. Durch ihn habe ich sehr viele neue Methoden, effizientes Arbeiten und vor allem viel über die Phänomenologie hinter den oft sehr technischen Rechnungen gelernt. Ein besonderer Dank auch dafür, dass ich mit Hilfe seiner Unterstützung an hochkarätigen Schulen wie z.b. in Cargese 2010 teilnehmen konnte, und mehrere Vorträge an internationalen Konferenzen halten durfte.

Ganz besonders bedanken möchte ich mich bei Simon Badger für die hervorragende Zusammenarbeit während der Doktorarbeit und auch darüber hinaus. Durch ihn habe ich viele Einblicke in das weite und faszinierende Feld der Unitaritätsmethoden bekommen. Für die gute Zusammenarbeit bedanke ich mich auch bei Valery Yundin, insbesondere für die vielen Computertricks mit denen er mir weitergeholfen hat. Einen speziellen Dank möchte ich an Theodor Schuster für die vielen fruchtbaren Diskussionen und die vertrauensvolle Zusammenarbeit richten.

Bedanken möchte ich mich auch bei der ganzen Arbeitsgruppe "Phänomenologie der Elementarteilchen" für die verschiedenen großen und kleinen Hilfen in der täglichen Arbeit, sei es durch Fachdiskussionen, Assistenz bei technischen Problemen, Rechentricks oder die auflockernden Gespräche in der Kaffeepause. Ein besonderer Dank geht in diesem Zusammenhang an Mohammad Assadsolimani, Peter Galler und Philipp Kant. Insbesondere ist es mir ein Anliegen, mich bei Bas Tausk zu bedanken, der sich immer Zeit für Fragen nahm, und dessen Quelle an Wissen über Teilchenphysik und Quantenfeldtheorie nie versiegte.

Besonders bedanken möchte ich mich bei Mohammad Assadsolimani, Stefanie Biedermann, Peter Galler, Lucia Gaschick, Bas Tausk, Louïse Vilén-Zürcher und Konstantin Wiegandt dafür, Teile dieser Arbeit Korrektur gelesen zu haben und damit entscheidend zum Gelingen beigetragen zu haben.

Des weiteren danke ich allen Verantwortlichen des Graduiertenkollegs 1504 "Masse Spektrum Symmetrie" der deutschen Forschungsgemeinschaft für die Finanzierung des größten Teils meiner Doktorarbeit, die vielen Reisen, die dadurch ohne weiteres möglich waren und für die interessanten Blockkurse und Vorlesungen.

Ganz ein besonderer Dank gilt all denen, die trotz großer Distanzen und Einbindung in ihre Berufe so oft und zuverlässig unsere Tochter Luise gehütet haben: Meine Mutter Gisela Biedermann, meine Schwiegermutter Ruth Stoll, meine Schwägerin Annette Stoll und Luises Taufpatin Anna Vogt. Ohne die viele Zeit und den freien Kopf, den sie mir geschenkt haben, würde die Arbeit nicht in dieser Form vorliegen. In diesem Zusammenhang gilt noch ein Mal mein besonderer Dank meinem Doktorvater Peter Uwer, der mir durch den Sonderforschungsbereich "SFB/TR9 Computergestützte Theoretische Teilchenphysik" einen Heimarbeitsplatz einrichten konnte, und mir durch viel Entgegenkommen eine sehr große Hilfestellung war, um Familie und Beruf zu vereinbaren.

Mein herzlichster und innigster Dank gilt Stefanie. Sie war und ist zusammen mit Luise die unerschöpfliche Quelle meiner Motivation.

Selbstständigkeitserklärung

Ich erkläre, dass ich die vorliegende Arbeit selbstständig und nur unter Verwendung der angegebenen Literatur und Hilfsmittel angefertigt habe.

Berlin, den 13. Juni 2013   Benedikt Georg Biedermann